W9-AQO-598

JUL 1971
RECEIVED
OHIO DOMINICAN
COLLEGE LIBRARY
COLUMBUS, OHIO
43219

DOOM OF A DREAM

DOOM OF A DREAM

Edited by

Doris R. Asmundsson

Queensborough Community
College

CHANDLER PUBLISHING COMPANY

An Intext Publisher • Scranton / London / Toronto

Library of Congress Catalog Card Number 76-140216
International Standard Book Number 0-8102-0414-2
Printed in the United States of America.

To Eric
and his generation

CONTENTS

IV. Cities

V. Population

VI. Future Prospects

PREFACE

Like the Moliere character who was surprised to discover that he had been speaking prose all his life, the American public seems to be amazed over the realization that it is surrounded by the environment. In the past few years this awareness has spread to the extent that ecology has become a familiar word and the average citizen, once securely isolated from nature, now expresses a concern for the natural world that was formerly limited to conservationists and bird watchers.

It is a lamentable irony that while we have been spending so many years and so many billions of dollars perfecting weapons that are now capable of destroying the entire world, we have at the same time unconsciously been contributing to the destruction of the world merely by living on it. An additional irony is the fact that although ecologists have issued dire warnings for years on the deterioration of the environment, it is only since man ventured into space and landed on the moon that he has begun to regard his own planet as something infinitely precious. From the cold, dead moon the astronauts looked back with longing to the earth and marveled over its beauty.

Only now, when we look at our strangled lakes and rivers, our crowded, murky cities, and our towering garbage dumps, do we realize with horror what we have done. And the reports of disaster continue to pile up. The forests outside Los Angeles are withering from clouds of carbon monoxide; a Greek ecology committee warns that Athens will have to be abandoned in ten years unless radical measures are taken to control air pollution; Thor Heyerdahl encountered lumps of tarlike material and large areas of filthy water in mid-Atlantic far from any shore; deaths from eating mercury-poisoned fish are reported daily. And there is no running away from the plague of pollution, no quick injection of serum, no handy technological cure-all. We ask in desperation, What can be done to save what has not yet been destroyed irrevocably?

Optimistic government officials and legislators seem to have unbounded faith in the passage of a scattering of weak laws which they hope will convince industrial polluters to control voluntarily their own pollution. Similar naïveté is expressed by those who believe that science will come up with some ingenious solution to all our pollution problems, thus saving us the trouble of modifying in any way the comfortable style of living to which we have become accustomed. In line with this train of thinking is the belief that no matter how serious environmental problems become, we must never attempt a solution which will in any way affect the economy of the nation. Health, sanity, and even life may have to be sacrificed, but never economic growth.

An excellent television program on the serious environmental problems that may exist in fifteen years was recently presented nationwide. It was a chilling story, but serious ecologists must have gotten an even greater shock when the program was interrupted at one point by commercials for a soft drink with nonreturnable bottles and, even worse, for an enzyme detergent! A furious letter to the station accusing it of double talk brought the following response from the network vice president:

> The problem is to rejuvenate our environment without destroying our economy. I hope it will be possible for us to eliminate pollution by techniques of advanced technology rather than to eliminate those products, i.e., automobiles, gas, electricity, detergents, etc., which have been the leading contributors to our economy.

It is this type of thinking that must be revolutionized before any real progress will be made in saving our world. When we come to realize that what is good for General Motors is fatal to our lungs, we will refuse to buy another internal combustion engine. The automobile manufacturers will then presumably get the message. If it is possible to land a man on the moon, it is certainly possible to produce an electrically powered car.

But most pollution problems are considerably more complex, and it is this complexity that makes the situation so frustrating. Other problems lend themselves to easier solution: We know that the solution to war is peace, whether or not we have the will to bring it about; we know that the probable solution to nuclear terror is de-escalation and wise diplomacy. Both of these problems can be approached with a reasonable expectation of eventual solution. But pollution in the world hourly worsens as the population soars; the plague is everywhere, and many attempted solutions only result in greater pollution or damage in some other form. To burn garbage is to poison the air; to dump it at sea is to destroy marine life; to bury dangerous atomic wastes is to pollute underground water and to cause earthquakes. The interconnectedness of all human activity is suddenly evident.

The problem is many-sided and will require a strength of character that we in our comfortable society have not felt the need for in recent years. Will we have the courage to abandon the internal combustion engine? Will we be able to endure some years of economic retrenchment during the conversion process? Will we willingly pay higher taxes to provide for adequate sewage disposal, garbage removal, and urban renewal of a meaningful kind? Will we be able to use and re-use water that may be unattractive in taste and color? Will we decide to cut down on our wasteful consumer habits and learn to live a simpler, but perhaps more rewarding, kind of life? And most important and basic to the solution of all our problems, will we cut down on our exploding population? Only when the importance of these sacrifices has been made clear to the majority of our people can we hope for any check to the steady decline of our environment. And while we hesitate, the world deteriorates. In Greek mythology, Sisyphus was forced to push uphill a giant boulder which constantly rolled back upon him. We are in a similar position: We are struggling to push the world out of a murky hollow to the clear air above while it rolls back on us, always more heavily burdened with the pollution of our own making.

Doris R. Asmundsson

what if a much of a which of a wind

what if a much of a which of a wind
gives the truth to summer's lie;
bloodies with dizzying leaves the sun
and yanks immortal stars awry?
Blow king to beggar and queen to seem
(blow friend to fiend:blow space to time)
—when skies are hanged and oceans drowned,
the single secret will still be man

what if a keen of a lean wind flays
screaming hills with sleet and snow:
strangles valleys by ropes of thing
and stifles forests in white ago?
Blow hope to terror;blow seeing to blind
(blow pity to envy and soul to mind)
—whose hearts are mountains,roots are trees,
it's they shall cry hello to the spring

what if a dawn of a doom of a dream
bites this universe in two,
peels forever out of his grave
and sprinkles nowhere with me and you?
Blow soon to never and never to twice
(blow life to isn't:blow death to was)
—all nothing's only our hugest home;
the most who die,the more we live

e. e. cummings

From *Poems 1923-1954*, by e. e. cummings (New York, 1944), p. 401.

PART I

ECOLOGY

From the Book of Genesis

In the beginning God created the heaven and the earth. And the earth was without form, and void; and darkness was upon the face of the deep. And the spirit of God moved upon the face of the waters. And God said, "Let there be light": and there was light. And God saw the light, that it was good: and God divided the light from the darkness. And God called the light Day, and the darkness he called Night. And the evening and the morning were the first day.

And God said, "Let there be a firmament in the midst of the waters, and let it divide the waters from the waters." And God made the firmament, and divided the waters which were under the firmament from the waters which were above the firmament: and it was so. And God called the firmament Heaven. And the evening and the morning were the second day.

And God said, "Let the waters under the heaven be gathered together unto one place, and let the dry land appear": and it was so. And God called the dry land Earth; and the gathering together of the waters called he Seas: and God saw that it was good. And God said, "Let the earth bring forth grass, the herb yielding seed, and the fruit tree yielding fruit after his kind, whose seed is in itself, upon the earth": and it was so. And the earth brought forth grass, and herb yielding seed after his kind, and the tree yielding fruit, whose seed was in itself, after his kind: and God saw that it was good. And the evening and the morning were the third day.

And God said, "Let there be lights in the firmament of the heaven to divide the day from the night; and let them be for signs, and for seasons, and for days, and years: and let them be for lights in the firmament of the heaven to give light upon the earth": and it was so. And God made two great lights; the greater light to rule the day, and the lesser light to rule the night: he made the stars also. And God set them in the firmament of the heaven to give light upon the earth, and to rule over the day and over the night, and to divide the light from the darkness: and God saw that it was good. And the evening and the morning were the fourth day.

And God said, "Let the waters bring forth abundantly the moving creature that hath life, and fowl that may fly above the earth in the open firmament of heaven." And God created great whales, and every living creature that moveth, which the waters brought forth abundantly, after their kind, and every winged fowl after his kind: and God saw that it was good. And God blessed them, saying, "Be fruitful, and multiply, and fill the waters in the seas, and let fowl multiply in the earth." And the evening and the morning were the fifth day.

And God said, "Let the earth bring forth the living creature after his kind, cattle, and creeping thing, and beast of the earth after his kind": and it was so. And God made the beast of the earth after his kind, and cattle after their kind, and every thing that creepeth upon the earth after his kind: and God saw that it was good.

And God said, "Let us make man in our image, after our likeness: and let them have dominion over the fish of the sea, and over the fowl of the air, and over the cattle, and over all the earth, and over every creeping thing that creepeth upon the earth." So God created man in his own image, in the image of God created he him; male and female created he them. And God blessed them, and God said unto them, "Be fruitful, and multiply, and replenish the earth, and subdue it: and have dominion over the fish of the sea, and over the fowl of the air, and over every living thing that moveth upon the earth."

And God said, "Behold, I have given you every herb bearing seed, which is upon the face of all the earth, and every tree, in the which is the fruit of a tree yielding seed; to you it shall be for meat. And to every beast of the earth, and to every fowl of the air, and to every thing that creepeth upon the earth, wherein there is life, I have given every green herb for meat": and it was so. And God saw every thing that he had made, and, behold, it was very good. And the evening and the morning were the sixth day.

Thus the heavens and the earth were finished, and all the host of them.

There Are No Islands Any More

Dear Islander, I envy you:
I'm very fond of islands, too;
And few the pleasures I have known
Which equaled being left alone.
Yet matters from without intrude
At times upon my solitude:
A forest fire, a dog run mad,
A neighbor stripped of all he had . . .

(The tidal wave devours the shore:
There *are* no islands any more.)

Edna St. Vincent Millay

From *There Are No Islands Any More*, by Edna St. Vincent Millay (New York and London, 1940), pp. 5-6. Two portions of the poem.

The Island Earth

Margaret Mead

In 1940 Edna St. Vincent Millay wrote a poem called "There Are No Islands Any More," which moved those who were involved in World War II very deeply. The theme, that nowhere on this planet could man flee from man and be safe, that war and its aftermath reached to the most remote islands, tugged at the imagination of those of us who were living through the most widespread war in history, a war that culminated in the horrors of Hiroshima. People stopped talking about finding themselves an island where life could be lived out in peace with nature, and those who were fond of quoting added, from Donne, "No man is an island, entire of itself. . . ." Islands as a daydream of escape went out, and casual acquaintances stopped asking to be taken along on my field trips. When islands were mentioned, it was their vulnerabilities that were spoken of: population growth in Mauritius and Samoa; Japan's awareness of the need for population control; the devastating volcanic eruption in Bali that destroyed a third of the arable land; the unwillingness of Java's population to leave their crowded island for a less crowded one. The emphasis continued to be on the theme, "no place to go, no hiding place down here." Islands pointed out the interconnectedness of men on earth and their mutual vulnerability to each other's homicidal and genocidal aims.

The emergence of Indonesia as a new nation—the fifth largest in the world—was all the more striking because this is a nation made up of 80 million people living on 3,000 islands, and people raised their eyebrows when Indonesia tried to extend the limits of sovereignty to include the inland waterways of her watery empire. Buckminster Fuller

From *Natural History*, 79:22 and 102 (January 1970).

designed a map—a diomaxion map—which showed the continents of the earth as an interconnected land mass. Islands were definitely out, a handicap in some way or other to full-scale continental living.

Then came NASA and the moon program, and finally the first breathtaking photographs of the earth from the moon. Mankind joined the astronauts in their willowy, eerie, unweighted walks on the moon and saw the earth in all its isolated diversity. Earth became an island in space. The earth seen from the moon was a whole in a new sense, no longer simulated by a globe, but seen whole. Scientist fathers conversing with their small sons found themselves confused because they were still earthbound looking toward the moon, while the children were on the moon looking back toward earth.

Besides these major transforming events—the sense of political and military vulnerability that grew up after World War II, and the specific change in perspective that has grown with the space program as the earth has become planet Earth—something else has been happening. Men everywhere are becoming conscious that this planet, like any small island, is interconnected in ways other than war and rumors of war. The spread of radioactive dust; the long journey of DDT from someone's rose garden to the shell-less eggs of unborn birds and the bones of unborn children; the new, resistant strains of venereal disease and malaria, which are robbing us of our recent conquest of these dangers; the knowledge that man's activities can alter the temperature of the earth, create storms of inestimable strength, pollute the oceans as well as the small lakes and streams that are dying throughout the civilized world: all have brought home to us that the earth is an island. Interconnected the peoples of the earth are—vulnerable to each other's weapons and no longer able to defend their frontiers and their children; vulnerable also to the acts of people half a world away, as they casually dump tanks of nuclear by-products into the sea depths, which no one has yet properly explored, or send clouds of pollution through the air. As those who love and protect the wilderness and try to save a part of it for man, and as those who see their main crop destroyed by the by-products of human intervention in agriculture or animal husbandry, so now the whole world is coming to realize the interconnectedness between the way men live and whether or not their children and their children's children will have a habitable world. Not war, but a plethora of man-made things—disposable, indestructible beer cans; too much industrial waste in the lakes and streams, from antibiotics designed to protect egg-laying fowls to pesticides designed to protect the orange crop—is threatening to strangle us, suffocate us, bury us in the debris and by-products of our technologically inventive and irresponsible age.

With this new realization, which is expressing itself in a hundred different ways, from government commissions and antipollution groups, to the American Association for the Advancement of Science's Committee on Science in the Promotion of Human Welfare, to the Scientists Institute for Public Information, to small committees in small New England towns, the debate goes on. . . . With this proliferation of public interest, those who have been fighting these battles for conservation, for protection, for soil rehabilitation, for reforestation, and those who have become more recently aware of the dangers of pollution, overpopulation, and overload of every facility are meeting and looking for new ways of stating their common interests. Words like *ecosystem*, the whole interacting system in which a change in any one variable —temperature, the number of fish or of fishermen, a factory built on the banks of a stream, or a florist's seed field five miles away—may change the whole system, and *biosphere*, the whole natural living system of the planet and its surrounding atmosphere, are coming into the vocabulary of the concerned all over the world. These terms come from the science of ecology, a science that, on the whole, took as its model a pond, a lake, or a marsh and, while allowing for interaction among every natural component, took little cognizance of man himself, except as an interfering factor. If we wanted to teach our children about ecosystems, the model we used was an aquarium, in which the delicate relationships between water, plants, and aquatic creatures had to be watched over and kept in balance.

Aquariums are indeed a fine teaching aid and will give children an idea of the balance of the natural world, especially the great mass of urban children who meet nature either in the form of a pet who has to be walked in the streets or provided with "kitty litter." But it is becoming increasingly clear that this model, over which the aquarium owner stands, like a god, presiding over a small glass tank heated by electricity (itself vulnerable to a power failure) is only a very partial model of what is happening to us. The child's aquarium is a model of a world almost totally dependent on man, but of which he is a spectator and protector, not an integral part.

If, from the science of ecology, we try to develop a new profession of those who stand guard over the environment, we stand in danger of still leaving man outside, to become an "environmental manager," a significant factor, but not a true part of the natural world. To the core subject of ecology, it is suggested that we add the human sciences to to train aspirant young environmental managers to deal with the problem. As new subject matters develop in the field of urbanization—ekistics, urban planning, urban design—there is an attempt to patch together from a number of disciplines a new whole, a science of the total ecosystem, into which man, somewhat grudgingly, is to be admitted.

I do not think this is the way to do it. We have had many decades of various interdisciplinary projects. Either they represent a coalition of different disciplinary interests, in which each defends his own territory, or we get new incorporative fields, like economics or public health, which manufacture their own psychology and educational theory to suit themselves and, in turn, become little empires defending their domains against contenders.

I believe that there is another way to develop the kind of specialists that we will need as public concern for our endangered planet and our starving millions mounts. And this is where islands come back again. What students need to learn if they are to think about environmental protection and development is about whole inhabited ecosystems: ecosystems in which man himself, the way he plants and reaps and disposes of waste, multiplies or stabilizes his population, is a *conscious* factor. Man has molded and changed his environment since he learned to make tools and control fire. But in those days, perhaps a million years ago, he was not conscious of what he did, of how population was related to food supply, of how killing the young or eating all the eggs or gathering plants before they seeded would limit his future. It was on islands that man first began to learn these things. If there were too many people, either some would be driven out into the uncharted seas or there would be civil war. Some method of population control had to be adopted. Younger sons were forbidden to marry and infants exposed to die. Islanders knew when the birds came to nest, when the fish came to spawn, how periodic hurricanes affected their harvests. On many small islands today, the harsh realities of a rapidly changing world are forcing the men away to work, leaving only women and children at home. It was on islands that men first learned that they themselves were part of an ecosystem, so it is perhaps not surprising that the religious system of the ancient Polynesians emphasized taboo, that things were forbidden in the nature of the system itself. Under taboo, if men made no missteps they lived safely, but they had to be continuously alert to the consequences of infringement of the order of nature and the order of social life.

We need to find ways to understand, to teach children, and to prepare young men and women for careers in our interconnected and endangered world. The forces of public opinion are being marshalled nationally and internationally. A great international conference, conspicuous for its level of cooperation among usually rivalrous United Nations specialized agencies, was held in Paris in 1968. A conference on biology as the history of the future, sponsored by the International Union of Biological Sciences, was held in Chichén Itzá, Mexico, in January, 1969. At the initiative of Sweden, a great United Nations conference is being prepared for 1972. We need to have a model that will make man—always active, seldom conscious, irresponsible throughout most of history—a conscious participant in the development of planet Earth.

The smallest islands of the earth are almost all in trouble, whether it be the islands of the Hebrides, fighting the British Parliament and paying no income taxes; the burgeoning population of Mauritius; the belligerent population of Anguilla; or the small Greek islands whose men must all go away to sea. Such islands, grievously resourceless, overpopulated, and dependent upon distant and outside money, can become our models and our training grounds for the new professions that are needed. As small children were once asked to build a model of Solomon's Temple in Sunday School, or of Egyptian pyramids in day school to understand ancient civilizations centered on man alone and reflecting his natural environment, we now need materials so that each child in a class may have an island to think about: its size, its shape, its location, its weather, its resources, the habits and skills and despairs and hopes of its inhabitants, and its dependence upon world markets and diplomatic decisions in which its people have no part. And for those older students who wish to make a career of the protection and development of the whole of man's environment, a year on an island, learning the language, mastering the intricacies of the interrelationships of its living population and all its plants and creatures, would be perfect preparation for thinking about wholes. We would not need to patch disciplines together in an uneasy truce; members of various specialized disciplines could first obtain a firm grounding in their own fields and then—with a year's field work on an island—learn to articulate that speciality into a whole.

Following in Darwin's footsteps, Harold Coolidge began the trek back to islands for inspiration when he took a whole group of scientists to Galápagos in January, 1964. But the Galápagos have no human beings on them. It is the inclusion of people and their purposes that is now our problem. Nor need we ask islands—often in dire straits—to contribute, yet gain nothing from what they teach us about our planet Earth. Each student could be asked to work on some real problem, urgent to the people themselves, and thus prepare himself for the kind of world role when, in the 1970's and 1980's man's survival will hang in the balance—and the generation now growing up will have the task of saving this planet as a habitable spot for their children and their children's children.

All About Ecology

William Murdoch and Joseph Connell

The public's awakening to the environmental crisis over the past few years has been remarkable. A recent Gallup Poll showed that every other American was concerned about the population problem. A questionnaire sent to about five hundred University of California freshmen asked which of twenty-five topics should be included in a general biology course for non-majors. The top four positions were: Human Population Problems (85%), Pollution (79%), Genetics (71.3%), and Ecology (66%).

The average citizen is at least getting to know the word ecology, even though his basic understanding of it may not be significantly increased. Not more than five years ago, we had to explain at length what an ecologist was. Recently when we have described ourselves as ecologists, we have been met with respectful nods of recognition.

A change has also occurred among ecologists themselves. Until recently the meetings of ecologists we attended were concerned with the esoterica of a "pure science," but now ecologists are haranguing each other on the necessity for ecologists to become involved in the "real world." We can expect that peripatetic "ecological experts" will soon join the ranks of governmental consultants jetting back and forth to the Capitol—thereby adding their quota to the pollution of the atmosphere. However, that will be a small price to pay if they succeed in clearing the air of the political verbiage that still passes for an environmental policy in Washington.

Concern about environment, of course, is not limited to the United States. The ecological crisis, by its nature, is basically an international

From *The Center Magazine*, III:56-63 (January 1970).

problem, so it seems likely that the ecologist as "expert" is here to stay. To some extent the present commotion about ecology arises from people climbing on the newest bandwagon. When the limits of ecological expertise become apparent, we must expect to lose a few passengers. But, if only because there is no alternative, the ecologist and the policymakers appear to be stuck with each other for some time to come.

While a growing awareness of the relevance of ecology must be welcomed, there are already misconceptions about it. Further, the traditional role of the expert in Washington predisposes the nation to a misuse of its ecologists. Take an example. A common lament of the socially conscious citizen is that though we have enough science and technology to put a man on the moon we cannot maintain a decent environment in the United States. The implicit premise here seems clear: the solution to our ecological crisis is technological. A logical extension of this argument is that, in this particular case, the ecologist is the appropriate "engineer" to resolve the crisis. This reflects the dominant American philosophy (which is sure to come up after every lecture on the environment) that the answer to most of our problems is technology and, in particular, that the answer to the problems raised by technology is more technology. Perhaps the most astounding example of this blind faith is the recent assurance issued by the government that the SST will not fly over the United States until the sonic boom problem is solved. The sonic boom "problem," of course, cannot be "solved." One job of the ecologist is to dispel this faith in technology.

To illustrate the environmental crisis, let us take two examples of how the growth of population, combined with the increasing sophistication of technology, has caused serious problems which planning and foresight could have prevented. Unfortunately, the fact is that no technological solutions applied to problems caused by increased population have ever taken into consideration the consequences to the environment.

The first example is the building of the Aswan High Dam on the upper Nile. Its purposes were laudable—to provide a regular supply of water for irrigation, to prevent disastrous floods, and to provide electrical power for a primitive society. Other effects, however, were simply not taken into account. The annual flood of the Nile had brought a supply of rich nutrients to the eastern Mediterranean Sea, renewing its fertility; fishermen had long depended upon this annual cycle. Since the Aswan Dam put an end to the annual flood with its load of nutrients, the annual bloom of phytoplankton in the eastern Mediterranean no longer occurs. Thus the food chain from phytoplankton to zoöplankton to fish has been broken; and the sardine fishery, once producing eighteen thousand tons per year (about half of the total fish catch), has dropped to about five hundred tons per year.

Another ecological effect of the dam has been the replacement of an intermittent flowing stream with a permanent stable lake. This has allowed aquatic snails to maintain large populations, whereas before the dam was built they had been reduced each year during the dry season. Because irrigation supports larger human populations, there are now many more people living close to these stable bodies of water. The problem here is that the snails serve as intermediate hosts of the larvae of a blood fluke. The larvae leave the snail and bore into humans, infecting the liver and other organs. This causes the disease called schistosomiasis. The species of snail which lives in stable water harbors a more virulent species of fluke than that found in another species of snail in running water. Thus the lake behind the Aswan Dam has increased both the incidence and virulence of schistosomiasis among the people of the upper Nile.

A second example we might cite is the effect of DDT on the environment. DDT is only slightly soluble in water, so is carried mainly on particles in the water for short distances until these settle out. But on tiny particles in the atmosphere it is carried great distances; it may even fall out more heavily in distant places than close to where it was sprayed. DDT is not readily broken down by microörganisms; it therefore persists in the environment for many years. It is very soluble in fats so that it is quickly taken up by organisms. Herbivores eat many times their own weight of plants; the DDT is not broken down but is accumulated in their bodies and becomes further concentrated when the herbivores are eaten by the carnivores. The result is that the species at the top of the food chain end up with high doses of it in their tissues. Evidence is beginning to show that certain species of predators, such as ospreys, are being wiped out as a result of physiological debilities which lead to reproductive failure, all caused by accumulations of DDT.

* * * *

Before pesticides were applied to North American spruce and balsam forests, pest populations exploded once every thirty years or so, ate all the leaves, and then their numbers plummeted. Since spraying began, the pests, in the absence of a balancing force of predators, are continually able to increase between sprayings. In two instances, in cotton fields in Peru and in cocoa plantations in Malaysia, the situation became so bad that spraying was stopped. The predators returned and the damage by pests was diminished to the former tolerable levels.

Another consequence of spraying has been that any member of the pest population which happens to be physiologically resistant to an insecticide survives and leaves offspring; thus resistant strains are evolved. Several hundred of these resistant strains have evolved in the last twenty years.

Because DDT is not present in concentrated form in the environment, it does not represent an energy resource common enough to support microörganisms. None has yet evolved the ability to break it down, even though it has been used as a pesticide for twenty-five years. Chlorinated hydrocarbons may even reduce drastically the plant productivity of the oceans. These plants are not only the base of the ocean food chain but also help maintain the oxygen supply of the atmosphere.

In sum, the indiscriminate use of DDT throughout the world, its dispersal by the atmosphere, its property of killing both pests and their enemies, and the evolution of resistant strains, have combined to create a crisis in the environment. The reaction has been to stop spraying some crops and to ban the use of DDT in some countries. Probably the correct solution, though, is to use pesticides carefully, applying them very locally (by hand if possible) to places where pest outbreaks are threatening, and to introduce or encourage enemies of the pests. This is called "integrated control." It is the hope of the future.

Since this article concerns pure ecology, it is probably worth distinguishing between pure and applied ecology. Applied ecologists are concerned with such problems as controlling pests and maximizing the yield from populations. Pure ecologists study interactions among individuals in a population of organisms, among populations, and between populations and their environments. (A population is a more or less defined group of organisms that belong to the same species.)

A brief indication of how some ecologists spend their time may be in order here. One of us (Connell) became interested in discovering what determines the distribution on the rocky seashore of a species of barnacle. He made frequent visits to the shore, photographed the positions of barnacles, counted their numbers at different levels on the shore at different life stages, noted the density and positions of predators, other barnacle species, and so forth. He developed hypotheses (in one area, that the limit to distribution is set by the presence of another barnacle species; in another, that beyond a certain height on the seashore a snail species eats them all) and tested the ideas by various experiments such as placing cages on the shore to exclude predators or removing the competing species. This work went on for several years and has now firmly established the two hypotheses.

Murdoch spent the past three years in the laboratory examining an idea about predators. The idea was that predators keep the numbers of their various prey species stable by attacking very heavily whichever species is most abundant. (The idea is a bit more complicated than that, but that is approximately it.) This entailed setting up experiments where different predators were offered different mixtures of two prey species at a variety of densities, and then counting the number eaten of each species. These experiments led to others, in order to test different sub-hypotheses. The conclusion was that predators would "switch" only under particular conditions.

* * * *

Ecologists face problems which make their task difficult and at times apparently insurmountable. It is a young science, probably not older than forty years; consequently, much of it is still descriptive. It deals with systems which are depressingly complex, affected by dozens of variables which may all interact in a very large number of ways. Rather than taking a census of them, these systems must be sampled. Ecology is one of the few disciplines in biology in which it is not clear that removing portions of the problem to the laboratory for experimentation is an appropriate technique. It may be that the necessary simplification this involves removes exactly the elements from the system which determine how it functions. Yet field experiments are difficult to do and usually hard to interpret. Ecology, moreover, is the only field of biology which is not simply a matter of applied physics and chemistry. The great advances in molecular biology resulted from physicists looking at biological systems (such as DNA), whose basic configuration is explicable in terms of the positions of atoms. But the individual or the population is the basic unit in ecology. It seems certain, then, that a direct extension of physics and chemistry will not help ecologists.

Finally, there is the problem that each ecological situation is different from every other one, with a history all its own; ecological systems, to use a mathematical analogy, are non-Markovian, which is to say that a knowledge of both the past and the present is necessary in order to predict the future. Unlike a great deal of physics, ecology is not independent of time or place. As a consequence, the discipline does not cast up broad generalizations. All this is not a complete list of the general problems ecologists face, but it may be enough to provide a feeling for the difficulty of the subject.

Ecologists, though, do have something to show for forty years' work. These are some of the general conclusions they have reached. . . .

•Populations of most species have negative feedback processes which keep their numbers within relatively narrow limits. If the species itself does not possess such features, or even if it does, the community in which it exists acts to regulate numbers, for example, through the action of predators. . . .

•The laws of physics lead to derivative statements in ecology. For example, the law that matter cycles through the ecosystem, to be used again and again. Or the law that energy from the sun is trapped by plants through photosynthesis, moves up the food chain to herbivores and then to carnivores as matter, losing energy at each successive conversion so that there is generally less energy and biomass in higher food levels than in lower ones. Ecologists have tried to take such truths from physics and construct more truly ecological generalities from them. Thus, to stay with the same example, it appears likely that there are never more than five links in any one chain of conversions from plant to top predator.

•It is probably true, on a given piece of the earth and provided that the climate doesn't change, that a "climax" ecosystem will develop which is characteristic of the area's particular features and that places with similar features will develop similar ecosystems if left undisturbed. Characteristically, a "succession" from rather simple and short-lived communities to more complex and more persistent communities will occur, though there may be a reduction in the complexity of the final community. We use "final" to mean that a characteristic community will be found there for many generations. We might go further and say that during the period of development disturbances of the community will result in its complexity being reduced. (Again, such statements will certainly arouse the dissent of some ecologists.)

•Finally, most ecologists would agree that complex communities are more stable than simple communities. This statement illustrates the difficulties faced by theoretical ecologists. Take some of its implications: What is complexity and what is stability in an ecological setting? Charles Elton embodied the idea in a simple, practical, and easily understood way. He argued that England should maintain the hedgerows between its fields because these were complex islands in a simple agricultural sea and contained a reservoir of insect and other predators which helped to keep down pest populations. . . . To keep the numbers of prey stable, the most likely candidates are predators. Now other questions arise: Do we just want more species of predators? Do we want more species of predators which are very specific in the prey they eat, implying that prey are stabilized by having many species feed on them? Do we want predators which are very general and attack many prey species, so that we still have a large number of interspecific interactions which are made up in a different way? The answer is not obvious, and indeed there is disagreement on it. Furthermore, if one studies the way some predators react to changes in the numbers of their prey, their short-term responses are such as to cause *instability*. Thus only some types of biological complexity may produce stability.

What do we mean by stability? In the examples cited, we have meant numerical constancy through time, but this is by no means the only meaning. It has even been suggested that numerical *in*constancy is a criterion for stability. Stability might also mean that the same species persist in the same area over long periods, showing the same sort of interspecific interactions (community stability). A community or population might be considered stable because it does not change in response to a great deal of environmental pressure, or because it changes but quickly returns to its original state when the disturbing force is removed. It is worth noting that if a population or community is observed merely not to change, we cannot tell whether this is owing to its ability to resist perturbing factors or merely to the absence of such factors. If we want to know about the *mechanisms* which might lead to the truth of our original statement, "complexity leads to stability," all the above points are important.

We submit that ecology as such probably cannot do what many people expect it to do; it cannot provide a set of "rules" of the kind needed to manage the environment. Nevertheless, ecologists have a great responsibility to help solve the crisis; the solution they offer should be founded on a basic "ecological attitude." Ecologists are likely to be aware of the consequences of environmental manipulation; possibly most important, they are ready to deal with the environmental problem since their basic ecological attitude is itself the solution to the problem. Interestingly enough, the supporting data do not generally come from our "abstract research" but from massive uncontrolled "experiments" done in the name of development.

These attitudes and data, plus obvious manifestations of physical laws, determine what the ecologist has to say on the problem and constitute what might be called environmental knowledge. Some examples of this knowledge follow, though this is not to be taken as an encapsulation of the ecologist's wisdom.

• Whatever is done to the environment is likely to have repercussions in other places and at other times. Because of the characteristic problems of ecology some of the effects are bound to be unpredictable in practice, if not in principle. Furthermore, because of the characteristic time-dependence problem, the effects may not be measurable for years—possibly not for decades.

•If man's actions are massive enough, drastic enough, or of the right sort, they will cause changes which are irreversible since the genetic material of extinct species cannot be reconstituted. Even if species are not driven to extinction, changes may occur in the ecosystem which prevent a recurrence of the events which produced the community. Such irreversible changes will almost always produce a simplification of the environment.

•The environment is finite and our non-renewable resources are finite. When the stocks run out we will have to recycle what we have used.

•The capacity of the environment to act as a sink for our total waste, to absorb it and recycle it so that it does not accumulate as pollution, is limited. In many instances, that limit has already been passed. It seems clear that when limits are passed, fairly gross effects occur, some of which are predictable, some of which are not. These effects result in significant alterations in environmental conditions (global weather, ocean productivity). Such changes are almost always bad since organisms have evolved and ecosystems have developed for existing conditions. We impose rates of change on the environment which are too great for biological systems to cope with.

•In such a finite world and under present conditions, an increasing population can only worsen matters. For a stationary population, an increase in the standard of living can only mean an increase in the use of limited resources, the destruction of the environment, and the choking of the environmental sinks.

There are two ways of attacking the environmental crisis. The first approach is technology; the second is to reverse the trends which got us into the crisis in the first place and to alter the structure of our society so that an equilibrium between human population and the capacities of the environment can be established.

There are three main dangers in a technological approach to the environmental crisis. The first threatens the environment in the short term, the second concerns ecologists themselves, and the third, which concerns the general public attitude, is a threat to the environment in the long term.

Our basic premise is that, by its nature, technology is a system for manufacturing the need for more technology. When this is combined with an economic system whose major goal is growth, the result is a society in which conspicuous production of garbage is the highest social virtue. If our premise is correct, it is unlikely we can solve our present problems by using technology. As an example, we might consider nuclear power plants as a "clean" alternative to which we can increasingly turn. But nuclear power plants inevitably produce radioactive waste; this problem will grow at an enormous rate, and we are not competent to handle it safely. In addition, a whole new set of problems arises when all these plants produce thermal pollution. Technology merely substitutes one sort of pollution for another.

There is a more subtle danger inherent in the technological approach. The automobile is a blight on Southern California's landscape. It might be thought that ecologists should concern themselves with encouraging the development of technology to cut down the emission of pollutants from the internal combustion engine. Yet that might only serve to give the public the impression that something is being done about the problem and that it can therefore confidently await its solution. Nothing significant could be accomplished in any case because the increasing number of cars ensures an undiminishing smog problem.

Tinkering with technology is essentially equivalent to oiling its wheels. The very act of making minor alterations, in order to placate the public, actually allows the general development of technology to proceed unhindered, only increasing the environmental problems it causes. This is what sociologists have called a "pseudo-event." That is, activities go on which give the appearance of tackling the problem; they will not, of course, solve it but only remove public pressure for a solution.

Tinkering also distracts the ecologist from his real job. It is the ecologist's job, as a general rule, to oppose growth and "progress." He cannot set about convincing the public of the correctness of this position if in the meantime he is putting his shoulder behind the wheel of technology. The political power system has a long tradition of buying off its critics, and the ecologist is liable to wind up perennially compromising his position, thereby merely slowing down slightly or redirecting the onslaught of technology.

The pressures on the ecologist to provide "tinkering" solutions will continue to be quite strong. Pleas for a change of values, for a change to a non-growth, equilibrium economy seem naive. The government, expecting sophistication from its "experts," will probably receive such advice coolly. Furthermore, ecologists themselves are painfully aware of how immature their science is and generally take every opportunity to cover up this fact with a cloud of obfuscating pseudo-sophistication. They delight in turning prosaic facts and ideas into esoteric jargon. Where possible, they embroider the structure with mathematics and the language of cybernetics and systems analysis, which is sometimes useful but frequently is merely confusing. Such sophistication is easily come by in suggesting technological solutions.

Finally, there is always the danger that in becoming a governmental consultant, the ecologist will aim his sights at the wrong target. The history of the Washington "expert" is that he is called in to make alterations in the model already decided upon by the policymakers. It would be interesting to know what proportion of scientific advice has ever produced a change in ends rather than in means. We suspect it is minute. But the ecologist ought not to concern himself with less than such a change; he must change the model itself.

* * * *

We do not believe that the ecologist has anything really new to say. His task, rather, is to inculcate in the government and the people basic ecological attitudes. The population must come, and very soon, to appreciate certain basic notions. For example: a finite world cannot support or withstand a continually expanding population and technology; there are limits to the capacity of environmental sinks; ecosystems are sets of interacting entities and there is no "treatment" which does not have "side effects" (e.g. the Aswan Dam); we cannot continually simplify systems and expect them to remain stable, and once they do become unstable there is a tendency for instability to increase with time. Each child should grow up knowing and understanding his place in the environment and the possible consequences of his interaction with it.

In short, the ecologist must convince the population that the only solution to the problem of growth is not to grow. This applies to population and, unless the population is declining, to its standard of living. It should be clear by now that "standard of living" is probably beginning to have an inverse relationship to the quality of life. An increase in the gross national product must be construed, from the ecological point of view, as disastrous. (The case of underdeveloped countries, of course, is different.)

We do not minimize the difficulties in changing the main driving force in life. The point of view of the ecologist, however, should be subversive; it has to be subversive or the ecologist will become merely subservient. Such a change in values and structure will have profound consequences. For example, economists, with a few notable exceptions, do not seem to have given any thought to the possibility or desirability of a stationary economy. Businessmen, and most economists, think that growth is good, stagnation or regression is bad. Can an equilibrium be set up with the environment in a system having this philosophy? The problem of converting to non-growth is present in socialist countries too, of course, but we must ask if corporate capitalism, by its nature, can accommodate such a change and still retain its major features. By contrast, if there are any ecological laws at all, we believe the ecologists' notion of the inevitablility of an equilibrium between man and the environment is such a law.

We would like to modify some details of this general stand. Especially after the necessary basic changes are put in motion, there *are* things ecologists as "experts" can do: some of them are sophisticated and others, in a very broad sense, may even be technological. Certainly, determining the "optimum" U.S. population will require sophisticated techniques. Ecologists, willy-nilly, will have to take a central role in advising on the management of the environment.

* * * *

The Biological Revolution

Paul R. Ehrlich

In the several billion years that life has existed on the earth no event has been as startling as the rise of the species *Homo sapiens* to its present position of prominence. A mere eight thousand years ago mankind—then numbering perhaps five million individuals, far fewer than the number of such a contemporary species as bison—was just one of many kinds of large mammals. But even then man's hunting and food-gathering way of life was causing substantial disturbance in the planetary ecology. There is substantial evidence that Pleistocene man in America brought about the extinction of seventy per cent of the land mammals of large size, such as mammoths, horses, and camels; in Africa, about a third of the megafauna of the land was wiped out. Furthermore, many ecologists attribute the great grasslands of the world to primitive man's use of fire.

About 6000 B.C. the first groups of men, living on the edge of the Fertile Crescent in western Asia, gave up the nomadic life and settled down to agriculture. This change in man's way of life may have been the most important single happening in the history of the earth. It started a trend toward security from hunger for mankind, and initiated an irregular but persistent decline in the death rate in the human population. It also marked the beginning of the potentially lethal disturbance by man of the ecological systems upon which his life depends. When man practices agriculture he arrests the natural processes of ecological change at an unstable midpoint. Much of the planetary environment

From *The Center Magazine*, II:28-31 (November 1969).

has already been severely damaged. Now its utter destruction is threatened. In the last century alone the percentage of the earth's land surface classified as desert and wasteland has more than doubled, increasing from less than ten per cent to over twenty-five per cent, largely because of farming and grazing. Now mechanization and the use of pesticides, herbicides, and inorganic nitrogen fertilizers are rapidly accelerating the destruction of the earth's ecosystems. Insecticides alone have the potential of destroying the planet as a habitat for civilized man.

The agricultural revolution has been going on for about eight thousand years now; until a few hundred years ago it was the major cause of decline in the death rate. Population growth is a result of the difference between the birth and death rates, and birth rates have remained relatively high. Agriculture, therefore, has been largely responsible for the spectacular growth of the human population. Virtually alone, it caused the hundred-fold increase from five million to five hundred million between 6000 B.C. and 1650 A.D. Since then other revolutions, industrial and biomedical, have been added to the agricultural revolution. They have all contributed to reducing death rates. The human condition has improved sufficiently since 1650 for a further increase in the population to more than seven hundred times its size at the start of the agricultural revolution. By the nineteen-sixties some 3.5 billion human beings were crowded onto "Spaceship Earth."

While man's population has been growing, his culture has been evolving. He has developed a vast array of techniques for modifying his environment and himself, but he has failed to develop ways of understanding, guiding, and controlling his newfound abilities. Indeed our growing ability to change ourselves and our environments is at the heart of what is being called the biological revolution. For instance:

•Man has been extremely successful at lowering the human death rate, but has made no significant effort to lower the birth rate. As a result the human population is now growing at a rate which will double it in about thirty-five years, and the population growth continues to outstrip mankind's ability to produce and distribute food in proper quantities and of proper quality. We now have between 1.5 and two billion people living on inadequate diets. That is, there are now more hungry people on earth than the total world population in 1875. Although calories are in short supply, the most serious problem is probably the shortage of high-quality protein.

•Agricultural technology has developed to the point where very high food yields per acre are attained under certain conditions, primarily in the temperate zone. But the ecological consequences of this technology are widely ignored. Furthermore, modern fishing technology and the escalation of pollution threaten to destroy the resources of the sea on which man depends heavily for the all-important protein component of his diet. These resources are not unlimited, contrary to what one often reads in the popular press. We may already have exceeded the annual sustainable yield, and even under the best possible conditions more than a few-fold increase would be difficult to obtain.

•Molecular biologists have uncovered many of the basic chemical mechanisms of life and their work may be put to broad practical use in the near future. It should, for instance, soon be possible to predetermine the sex of a child and to correct certain inborn defects of metabolism. Indeed, the future potential for "genetic engineering" seems incredible. But then so does our lack of consideration of just what kinds of human beings we want to engineer, and to what purpose. The discoveries of molecular biology also pose a direct threat to human survival. Some molecular biologists are at work in chemical and biological warfare laboratories, engineering ever-more-lethal strains of viruses and bacteria which could very well bring the population explosion to an end once and for all.

•Medical scientists in the United States have followed the flow of federal money and, with varying degrees of success, put a great deal of effort into curing the kinds of disease suffered by middle-aged congressmen. Thus we have a spectacle of vast resources poured into programs leading to heart transplants for a very few individuals in a country where many millions are malnourished. The United States, furthermore, ranks only fifteenth in the world in infant mortality. It is, on the other hand, fortunate that the serious ethical problems associated with organ transplants and prostheses are being aired now while they are still a minor sideshow as far as the mass of humanity is concerned. For in the unlikely event that mankind should solve its pressing problems, reduce the size of the human population, and preserve a world in which medical science flourishes, these questions will become more serious than anything contemplated today. The most elemental questions will be: "what is an individual?" and "how long should an individual's life be preserved?" I can see no theoretical barrier standing in the way of our eventually achieving individual life spans of hundreds or even thousands of years, even though we have not yet made any significant progress in this direction. (We have increased the average life span, permitting more people to live out what is probably a genetically determined span. But there is no known reason why we should not discover how greatly to expand that span.) It is quite clear that in the near future the problem of rejection of transplanted foreign tissues will be more or less solved and substantial life extensions by transplants will be possible. But where will the replacements come from, who will pay for them, and who will decide on the allocation of parts in short supply?

•The most revolutionary of all man's prosthetic devices is probably the computer, which may be used as a replacement for, or an extension of, the human mind. Computers, in conjunction with modern communications systems, have already revolutionized the lives of people in the developed countries. They have done much more than facilitate the obvious breakthroughs in science, technology, and social science. Computers have changed the power structures of institutions from universities to governments; to some degree they have taken decision-making away from human beings. Indeed, there is now talk that technological advances in armaments may require such rapid reaction times that computers will have to make the decisions about whether we will or will not go to war.

•Man has begun, for the first time, to turn systematically toward the frontiers of the mind. At a strictly empirical level, so-called "brainwashing" has demonstrated the kinds of horrors possible. Holistic experiments on the mind, using drugs, hypnotism, and electrical and surgical intervention are being made increasingly in order to "change minds." Computers enter the picture here also. They have, for instance, been used successfully to teach children to read and write. Slowly but surely, biologists are beginning to unravel the secrets of the nervous system and learning the bases of perception and memory. It seems a safe assumption that various kinds of controlled biochemical manipulation of the mind eventually will be possible. Such manipulation could, of course, be used for what almost everyone considers an obvious "good"—for example, the cure of mental illness or retardation. However, the potential for misuse of this power, whether accidental or intentional, needs no elaboration.

From these few examples, we can see that because of biological revolutions we are confronted with a set of extraordinarily difficult social and political problems. Such problems are growing at an incredible rate. At the root of all of them is an increasingly efficacious biological technology, which had its origins in agriculture. (If man had never practiced agriculture it is unlikely that he would ever have practiced molecular biology.)

There is a tendency, in meeting new challenges, to solve the problems accompanying biological technology by encouraging the further growth of biological technology, without any careful consideration of the consequences of such a growth. Thus we see the further development of ecologically naive agriculture technology as a "solution" to the population-food crisis. How do we solve the problem of too many people? Develop a better contraceptive technology, but neglect critical questions of human attitudes toward reproduction. Shortage of organs for transplant? Grow them in tissue culture or develop artificial organs. Information overload? Build bigger and better computers and communications networks.

The questions that need asking are all too rarely raised: What for? What kind of life are those additional people we feed going to live? What will the composite men of the Age of Transplants do with their extra years of life? When we can "improve" our minds genetically or biochemically, what kind of world will we have to think about? What kind of information will flow through our improved communications networks, and be processed by future generations of computers? Is Western cultural evolution taking us where we want to go—and taking the rest of humanity where it wants to go? These are some of the fundamental questions raised by the biological revolutions of the last eight thousand years. As the pace of change accelerates, our chances of answering them satisfactorily and modifying our behavior are diminishing rapidly. It is possible that the rapid growth of technology will lead to that common end of runaway evolutionary trends—extinction. The signs now point that way.

PART II

AIR

The Mouth of the Hudson

Chemical air
sweeps in from New Jersey,
and smells of coffee.
Across the river,
ledges of suburban factories tan
in the sulphur-yellow sun
of the unforgivable landscape.

Robert Lowell

From *A Controversy of Poets: An Anthology of Contemporary American Poets*, edited by P. Leary and R. Kelly (New York, 1965), pp. 247-248. A portion of the poem.

Nineteenth-Century Industrialism

Charles Dickens

In all their journeying, they had never longed so ardently, they had never so pined and wearied, for the freedom of pure air and open country, as now. No, not even on that memorable morning, when, deserting their old home, they abandoned themselves to the mercies of a strange world, and left all the dumb and senseless things they had known and loved, behind—not even then, had they so yearned for the fresh solitudes of wood, hillside, and field, as now; when the noise and dirt and vapour of the great manufacturing town, reeking with lean misery and hungry wretchedness, hemmed them in on every side, and seemed to shut out hope, and render escape impossible.

"Two days and nights!" thought the child. "He said two days and nights we should have to spend among such scenes as these. Oh! if we live to reach the country once again, if we get clear of these dreadful places, though it is only to lie down and die, with what a grateful heart I shall thank God for so much mercy!"

With thoughts like this, and with some vague design of travelling to a great distance among streams and mountains, where only very poor and simple people lived, and where they might maintain themselves by very humble helping work in farms, free from such terrors as that from which they fled—the child, with no resource but the poor man's gift, and no encouragement but that which flowed from her own heart, and its sense of the truth and right of what she did, nerved herself to this last journey and boldly pursued her task.

From *The Old Curiosity Shop*, by Charles Dickens (London, 1911), pp. 323-327. Reproduced here under the editor's title.

"We shall be very slow to-day, dear," she said, as they toiled painfully through the streets; "my feet are sore, and I have pains in all my limbs from the wet of yesterday. I saw that he looked at us and thought of that, when he said how long we should be upon the road."

"It was a dreary way he told us of," returned her grandfather, piteously. "Is there no other road? Will you not let me go some other way than this?"

"Places lie beyond these," said the child, firmly, "where we may live in peace, and be tempted to do no harm. We will take the road that promises to have that end, and we would not turn out of it, if it were a hundred times worse than our fears lead us to expect. We would not, dear, would we?"

"No," replied the old man, wavering in his voice, no less than in his manner. "No. Let us go on. I am ready. I am quite ready, Nell."

The child walked with more difficulty than she had led her companion to expect, for the pains that racked her joints were of no common severity, and every exertion increased them. But they wrung from her no complaint, or look of suffering; and, though the two travellers proceeded very slowly, they did proceed; and clearing the town in course of time, began to feel that they were fairly on their way.

A long suburb of red brick houses,—some with patches of garden ground, where coal-dust and factory smoke darkened the shrinking leaves, and coarse rank flowers; and where the struggling vegetation sickened and sank under the hot breath of kiln and furnace, making them by its presence seem yet more blighting and unwholesome than in the town itself,—a long, flat, straggling suburb passed, they came by slow degrees upon a cheerless region, where not a blade of grass was seen to grow; where not a bud put forth its promise in the spring; where nothing green could live but on the surface of the stagnant pools, which here and there lay idly sweltering by the black roadside.

Advancing more and more into the shadow of this mournful place, its dark depressing influence stole upon their spirits, and filled them with a dismal gloom. On every side, and as far as the eye could see into the heavy distance, tall chimneys, crowding on each other, and presenting that endless repetition of the same dull, ugly form, which is the horror of oppressive dreams, poured out their plague of smoke, obscured the light, and made foul the melancholy air. On mounds of ashes by the wayside, sheltered only by a few rough boards, or rotten pent-house roofs, strange engines spun and writhed like tortured creatures; clanking their iron chains, shrieking in their rapid whirl from time to time as though in torment unendurable, and making the ground tremble with their agonies. Dismantled houses here and there appeared, tottering to the earth, propped up by fragments of others that had fallen down, unroofed, windowlesss, blackened, desolate, but yet inhabited. Men, women, children, wan in their looks and ragged in attire, tended the engines, fed their tributary fires, begged upon the road, or scowled half-naked from the doorless houses. Then came more of the wrathful monsters, whose like they almost seemed to be in their wildness and their untamed air, screeching and turning round and round again; and still, before, behind, and to the right and left, was the same interminable perspective of brick towers, never ceasing in their black vomit, blasting all things living or inanimate, shutting out the face of day, and closing in on all these horrors with a dense dark cloud.

Man and His Home

A. J. Haagen-Smit

Earth finally made it. It, too, had its day. It took some doing, but finally the younger generation got up in arms and organized sit-ins and teach-ins—all for good old Mother Earth. It was about time. The airplane commuter knows when he is nearing our big cities—haze and brown clouds are the rule. The views from the Empire State Building in New York, the Prudential Building in Chicago, the Humble Building in Houston, or the City Hall in Los Angeles have one thing in common. Visibility is considered good when one can see the outlines of hazy streets below and a horizon covered by a brownish haze.

The waters from the sewers of the big cities and their industries stretch their greenish and brownish effluent far into the lakes, and we read about the replacement of healthy fish populations by mudsuckers and unsightly fields of algae, choking all other plant life to death. Excessive human deaths during periods of heavy air pollution have been recorded in London, New York, and other cities.

Today, a majority of Americans, urban and rural alike, in all sections of the country live near polluted waters and are breathing polluted air. In some cases, smog is doing more damage to crops than even insects. We have upset the balance of nature because we have assumed that our resources—air, water, and soil—are infinite.

This century has taught us that our space ship is not at all so big and that we have finally succeeded in changing, on a global scale, the

From *Vital Speeches*, XXXVI:572-576 (July 1, 1970). Presented in cooperation with the American Society of Landscape Architects at its annual meeting, Williamsburg, Virginia, April 28, 1970.

composition of our atmosphere and the water of the oceans. The rapid burning of fossil fuel in the last fifty years has raised the concentration of carbon dioxide in the air substantially, and this, together with particulate matter from our industrial operations, will affect the heat balance of the earth.

All these terrifying incidents have brought home to us the realization that we are part of Nature, that there is a close bond between us and our environment. In the past, it was taken for granted that we were the masters of all living and inanimate things on earth and that we could take care of the needs of man forever. The general thought, if there was any, was: the soil, the water, and the air are ours, and no one is going to tell us what we are going to do with them. This thought still prevails in the world, but there are encouraging signs that we have reached a more mature consideration of the environmental problems that this Nation and the world face.

Whenever we pick up a newspaper, we seldom fail to see an article on the new science:

Actually, this is not a new field of science at all. It is the study of the *home*, or, rather, the environment. The name comes from *Oikos*, meaning house, and *Logos*, which is discourse. Ecology, then, is the *Study of our House*, or, in a broader sense, our environment. This may be the total biosphere of the earth, continent, or smaller national units, such as forests, islands, or even the small world of a square foot of soil. This discipline was practiced by a rather small group of scientists, often pictured as collectors of butterflies and shells. Actually, these scientists have a true and profound curiosity—simply for the sake of knowing—about everything that goes on in nature.

They are the ones that know about the delicate balance between the living world and the environment. If we had listened to their teachings, we might have prevented many of our predicaments. We could have learned that we, too, are a part of the total systems of living and nonliving things, and that a thorough knowledge of the functioning of our *house*, or *environment*, is fundamental to our survival. Many civilizations of the past have succumbed because of a lack of understanding of the laws of nature. These laws cannot be flouted for long without severe punishment. Exhaustion of the soil and slash and burn techniques led to the downfall of early empires of the Mayas in Mexico and the Persians in Western Asia.

After finding out what substances were in the air, I could not help but get interested in their origin—where they came from and how to prevent their emission. The combination of chemical and community problems is fascinating, and I was drawn deeper and deeper into environmental pollution studies.

* * * *

Our ancestors lived in the happy certainty that the earth was infinite and that there was enough soil, water, and air to go around. The system was well balanced, in a steady state of equilibrium that was not going to change much during their presence on earth. It was realized, of course, that there are some changes, some violent upheavals—storms, floods, earthquakes, births, and deaths—but they form a recurring pattern and we get used to such a situation. The whole system seemed to have a kind of comfortable stability.

Looking at an astronaut's view of the earth, we begin to realize that the earth is actually not so large at all, and that the stability applies only to our time scale of geological and evolutionary happenings. A continuous flow of events led from the origin of life some two to three billion years ago to the elaborate structures that we represent. For evolutionary processes, *changes* in the environment were essential; however, for the continuation of the species, the *constancy* of the environment was of great importance. Even small changes in the environment will eventually lead to the disappearance of the species, or its replacement by others more suitable to new conditions.

When organisms first climbed on land some half a billion years ago, they soon found out that the scorching sun in the daytime and the cold of the night were quite different from the protective life that they had left. And the all-pervasive supply of food in the ocean was replaced by localized supplies for which they had to hunt. When the organisms had acquired the ability to use the sun's energy, when photosynthesis had evolved, it gave them a new freedom to move without being dependent on the nutrients dissolved in the oceans. It was a rough world for the adventurous mutants, but those that had established themselves survived in the uninhabited areas. Evolution and adaptation to the changed world went on, and today there are one to two million successful trials known—350,000 in the plant world, and at least more than one million in the animal world.

Most remarkable are the methods used to overcome the limitations set by the environment. Some of the descendants of a green alga, for example, formed an alliance with another organism, a fungus, and together they survived the new situation. The alga lives inside the fungus, protected from the fluctuating humidity, while the fungus profits from the photosynthetic ability of the alga, which supplies it with nutrients. This combination, the lichen, was so successful that it is found in the most inaccessible and unpromising spots—on the bare rock in the deserts, as well as in the coldest regions of earth.

The less adventurous forms of life gave up their relatively safe surroundings only hesitatingly, and even today some of the present land animals, such as frogs, return to the water to insure their offspring of a better chance at life. Others found their safety in the dark areas under the surface of the earth, where the absence of large variations of light and humidity guaranteed their continued existence. Because of the stability of the surroundings, early forms of evolution found a more permanent environment in the dark and damp soil, and it now houses probably the largest supply of species, in numbers and in kind, anywhere in the world.

Actually, this soil cover shelters more life than can be found in any other stratum of any other environment on earth. The inhabitants exist in numbers that stagger the imagination.

Some years ago, some scientists blocked off a small section of forest soil in New York State and removed the top layer of earth to a depth of one inch. They made a careful count of the insects and other invertebrates found in one square foot of this top layer. In all, there was an average of some 1,400 living creatures, including 865 mites, 265 springtails, 22 millipedes, 19 adult beetles, and various numbers of 12 other forms.

Had an estimate also been made of the microscopic populations, it might have ranged up to 2 billion bacteria and many millions of fungi, protozoa, and algae in a mere teaspoonful of soil.

This underworld plays an important role in the development of higher plants. It prepares the soil, the humus, in which the plants sink their roots, where they find predigested foods made by bacteria, molds, insects, nematodes, and numerous other small animals. Were it not for the work of the soil creatures, the forest would soon be choked in its own waste, and vegetation would not be possible.

The bed of mulch is rich in nutrients; a seed carried by the wind will find a place to send its roots down, and a new life emerges – maybe a violet, maybe a pine tree.

We can say that wherever there was room, some empty niche, some organism found a place to live.

The study of the adaptation of species of plants and animals to take advantage of special circumstances is most fascinating. In the family of the African violet, for example, we find a plant that dips its roots in a rain-filled vase formed from one of its leaves. Other species of the same family spread such an unpleasant odor of decomposing meat that only a certain type of fly will visit it in preference to other flowers, which to us have a more pleasant smell.

The work of the scientist whom we honor today abounds with examples of the extraordinary inventiveness of the living world to take advantage of opportunities in the environment.

In the animal world it is no different. Think of the remarkable homes the termites build—which may be twenty or more feet high—to overcome the limitations set by nature. The termites create their own optimum environment and are first-rate air conditioning engineers. Their homes are temperature-controlled and ventilated to keep the carbon dioxide concentration down. They also are expert gardeners, and inside the mound they cultivate fungus gardens to supply the queen with nutrients.

It took the earth a few hundred million years to establish its lush vegetation and establish the concentration of the major constitutents of our present-day atmosphere. In the photosynthetic processes of plants, carbon dioxide was converted to complex organic materials, and large quantities of oxygen were released into the atmosphere. In the course of millions of years, huge quantities of chemical energy were stored in these organic materials. They represent nature's stores of fossil fuels, coal, shale, oil, and gas. It was during that time that the composition of the atmosphere reached 21 per cent—or 210,000 parts per million—of oxygen, and .03 per cent—or 300 parts per million—of carbon dioxide. The living world had time to adjust itself to the gradual change from a reducing to an oxidizing atmosphere.

And, in the course of time, species came and went; some lasted only a few million years; others, such as dinosaurs, survived some 150 million years. Their contemporaries, the gingko trees and the cycads in our gardens, are still growing. They were witnesses of the struggle for existence and for power of a succession of masters of the world.

And then, about a million years ago, *man came.* His early existence must have been a precarious one. His home, probably a hole in the rocks, was all he had to protect himself from the rough environment and his enemies. His ability to master the art of making fire, and later the exploitation of the fossil fuel supplies, freed hands and freed minds to think.

Living conditions improved, diseases were conquered, man prospered, and his numbers increased rapidly. The cave became a village, the village a town, and today the towns have melted together in a new form, the "megalopolis."

It is estimated that up to about the birth of Christ, there were only two people per square mile of earth's surface. Today there are about a hundred, and by the year 2000 this number will be doubled. Such a calculation, however, gives only part of the picture. People are not evenly spread over the face of the earth. On the contrary, in urban areas the density of our present-day population has to be counted in tens of thousands per square mile.

It was the industrial revolution—the use of energy from coal and fossil fuel in general—that made this population growth possible. It was like having an army of slave workers. These modern slaves are calories or kilowatt hours or British Thermal Units. The amount of energy available to a single person, expressed in human labor, would correspond to the work of a hundred slaves. It is like Aladdin's lamp. A simple rub and there appear the slaves. A simple turn of your key in the car and several hundred horsepower, corresponding to a thousand slaves, spring into action.

This is, of course, wonderful, but the trouble is that the energy slaves are not very neat. In the process of burning our fuels we use up oxygen, but—what is more objectionable—we also add small amounts of toxic material to the air. Soot and sulfureous fumes became, in the historic times, the attributes of the devil.

Pollution of the air disturbed one of the kings of England in the Fourteenth Century so much that the use of a certain type of coal was forbidden. Infringers of the rule were fined and their ovens demolished in case of repetition. One unfortunate individual was condemned to death because he had infringed on the smoke rule three times.

Some 400 years later, Joseph Priestley discovered the essential, life-giving element of the air, *oxygen,* and he made the prophetic remark: "Who can tell but in time this pure air may become a fashionable article and luxury, hitherto only two mice and myself have had this pleasure, privilege of breathing it. It may be peculiarly salutory to the lungs in certain morbid cases when the common air would not be sufficient to carry off the *phlogistic putrid effluvium* fast enough."

These remarks were undoubtedly inspired by the heavy pollution in the industrial towns of England, and his home town, Birmingham, was just as bad as London with its black fogs.

Something had happened, and what it was I like to illustrate with the nostalgic writings of Chateaubriand, the French ambassador to England, upon a return visit to London in the beginning of the Nineteenth Century: "I have seen England with its old customs and its old prosperity, the small and lonely church with it tower, the cemetery, its small streets and the heather dotted with sheep. *Where is it now?*

"No more woods, less birds, less fresh air. Today its valleys are obscured by the fumes of *smelters and factories.* Oxford and Cambridge take on a look of ghost towns, their colleges and Gothic chapels are half abandoned. In their cloisters among the graves of the past lie, forgotten, the marble annals of people from long ago. *Ruins guarding ruins.*"

Those blackened relics stand as the tragic sins of the new era, the industrial revolution, with its thoughtless use and mismanagement of our natural resources. *A disregard of the most elementary right—the right to breathe clean air.*

This, too, is the story that runs through the history of American municipalities. It is one of rapid growth in population and industrial activity, marked by wastefulness of material resources, carelessness in regard to the future, indifference to many things of life, and a blind opposition toward anything that seems to threaten, in even a remote way, that which is termed prosperity.

A panoramic view of most of our large cities as seen from their skyscrapers shows a grayish and brownish fog limiting visibility to only a few miles. This is seen in New York, as well as in Los Angeles, London, or Yokohama, to name only a few of the affected areas. Air pollution has become a normal aspect of urban living, and it has brought with it irritation of the throat, nose, and eyes, and in some instances has caused death.

The end of 1967 witnessed an exciting event when the population counter went to 200 million. The press was jubilant about the accomplishment: more people meant more business, more cars sold, more building, and more advertising. To many this was prosperity; to others it was a day of gloom, a day that shows with deadly accuracy what our fate will be many years ahead. The population curve is going up without wavering.

Every 7½ seconds a baby is born; every 17 seconds someone dies. This means five more persons every minute, or 300 per hour, 7,200 per day. There are no cease-birth agreements, no holidays in this business. In one year there are some two million more mouths to feed—a line of baby carriages stretching from New York to Los Angeles. With computer accuracy, we will celebrate the 300 million mark in only a few decades. Eighty-five per cent of these people are concentrated in the big cities, and most of them are subject to some degree of air pollution. Burning of fuels adds particulate matter in the form of soot or ash, and gases such as carbon dioxide, carbon monoxide, and oxides of nitrogen and of sulfur.

The concentration of people in cities has affected the meteorology or the climate within the built-up area. With many combustions going on, with a decreased wind circulation, and with a poor reflection of the sunlight, the temperature inside the cities is raised considerably over the surrounding rural area. This increase in temperature may be several degrees Fahrenheit. An interesting by-product of this rise in temperature is air circulation driving pollutants towards the center of the city.

Due to the polluted atmosphere, the solar energy received by the area may be in the order of 20 per cent, and a loss of half of the visible radiation and two-thirds of the ultraviolet radiation is not rare at all.

Atmospheric and terrain conditions that lead to a lack in ventilation occur far more frequently than most people realize, and it is not strictly necessary to have mountains to obstruct the flow of air. The streets in our metropolitan areas act as small canyons, and on windless days relatively high concentrations of pollutants are found there.

The pollution problems in our cities are aggravated by other larger scale meteorological phenomena. In many areas, especially on the Pacific Coast, the sinking or subsidence of air causes it to heat up slightly, and a condition is established by which the warm air is lying on top of a colder ground layer. This type of inversion layer ranges from a height of a few hundred feet to a few thousand. In other areas the earth radiation of heat during the night causes strong ground inversion. In both cases, pollutants caught in this colder layer refuse to rise and, consequently, create air problems.

It has been established that this existence of an inversion base within 500 feet of the surface occurs on more than 50 per cent of the nights in a year over most of the United States.

Under these conditions of limited ventilation, pollutants are held to the ground and are especially bothersome.

The effects of the pollutants are many: irritation of the throat and eyes, haze, and damage to vegetation are frequent. In addition, there is a quite serious effect that does not have the glamour of some of the more dramatic impressions of pollution, and that is the *damage to materials.* It is difficult to assess the value of the damage done, and the cost of repairing the damage; but it is certain to run into many billions of dollars per year. In a number of instances the damage can never be undone.

The simultaneous oxidative and reductive processes, or alkaline and acid atmospheres, do not respect any material, be it stone, paint, cloth, or various metals. Since the industrial revolution, art works have suffered irreversible damage. A frieze on the Parthenon in Athens, from which a plaster cast was made in 1802, shows the relatively minor damage that occurred during its first 2,240 years. A photograph of the same marble taken in 1938 is almost unrecognizable because of the rapid deterioration during the intervening 136 years of the industrial age. Building materials nowadays have to be chosen so as to withstand the onslaught of population. Deposits of chemicals eat their way in the smooth surfaces of metals, acting directly by setting up tiny electrochemical cells to pit and mar the surface.

Stones are corroded by normal weathering accelerated by pollutants; smoke and tarry deposits adhere to buildings and produce the unsightly and depressing look of an old city.

Oxides of sulfur, common pollutants of air in all urban communities, cause deterioration of metals and building materials. They attack marble and carbonate-containing stone; corrosion of building stone statuary is common all over the world; and in many cities conservationists are moving works of art indoors to protect them from the action of acidic pollutants. It is reported that Cleopatra's Needle has deteriorated more since its arrival in New York in 1881 than it did during three thousand years spent in Egypt.

Calcareous materials such as limestone, marble, lime plaster walls, and frescoes are subject to chemical assault by the sulfuric acid formed from moisture and sulfur oxides. The calcium carbonate in the stone converts to calcium sulphate and its hydrated form, gypsum, both of which are water soluble. In the process, the volume of the stone expands by 70 per cent. Stress and leaching result, and ultimately the stone crumbles. Granites and sandstones are not similarly affected.

J. V. Noble, from the Metropolitan Museum, explicitly implicated air pollutants when he wrote:

> The presence of various forms of sulfur in the air is particularly injurious to limestone and marble. There is an appreciable, visible etching on marble . . . I would say that all the exposed stonework of ancient elements at the Cloisters has deteriorated since its erection in New York City as a direct result of air pollution . . . It is pointless to collect outstanding works of art, many over a thousand years of age, if one thousand years from now they are going to be so badly deteriorated to be virtually worthless.

The situation in a museum which houses a collection of objects valued for their artistic or historical value presents a problem quite different from that of protecting consumer goods. In this case, the materials exposed to the atmosphere represent the whole spectrum of products used by men, from plant, animal, or mineral origin, natural and synthetic dyes, fabrics and paint, metals and plastics.

Libraries, as well as art museums, have concern for the effects of air pollution. The same acid atmosphere in New York City that decomposes the Egyptian statuary outside the Metropolitan Museum of Art also decomposes the pages of old books stored in the City's Central Research Library on Fifth Avenue. E. G. Freehafer, Director of the New York City Public Library, has estimated that about 1.8 million of the 4.3 million volumes in the Central Research Library are in an advanced state of deterioration. Air pollution is a prime factor in this problem, and the City Library has spent $900,000 since 1952 to microfilm decaying books.

The sources are well known: too many cars, too much industry, and too many people. Federal, State, and local agencies are all engaged in air pollution control. Severe restrictions have been placed on industries, and a strict program of automobile control is underway. By 1975, the new cars will be largely controlled. There is still, of course, the large number of older, partially controlled vehicles, but there is all reason to expect that methods will be found to reduce their emissions, too.

Even though the control authorities are doing a job unequaled in the world, the trouble they are facing is the explosive growth. Every year there is the equivalent of one more city of nearly two million people, plus their industry, plus their cars, to control, when we have not even caught up with the old pollution. This is a gigantic task.

But while some of us are trying to master the air pollution problems, other trouble spots appear. New problems arise in the demand for more products from our land. Intensive cultivation demanded the use of insecticides, herbicides, and nematocides, but because of careless use of such substances DDT has turned up in penguins in the South Polar regions and I am sure, in us, too. And a modern version of the children's story sounds like this: "The bird eats the fish that ate the plant that ate the DDT that man made." It has upset the soil, flora, and fauna, which we have seen are essential for the preparation of the environment of the roots of higher plants. To the chemist, DDT is only 2-dichlorodiphenyltrichloroethane, and he is proud that he has found a more toxic chemical than any before. He probably never asked what happened to the use that is made of his synthesis. Recent experiments have shown that the toxic chemicals destroy part of the *underground* workers. The recovery is a slow one, taking sometimes a year or more.

It has become clear that, for the future, we shall have to do more than follow a system of repair. Unpopular as it may be, the further expansion of our city has to be planned with the avoidance of further air pollution in mind. Air-polluting industries may have to be located elsewhere. A revision of our Nineteenth Century thinking on transportation is in order. Many things can be thought of by laymen, but the *constructive* thinking by experts in city planning and city government is needed to come up with a plan that the community will *buy.*

The new deal in conservation of our resources is a *planned and preventive conservation.* Planned because water, air, and soil belong together. An air pollution problem is not solved by dumping the effluent in our rivers or estuaries. A water pollution problem is not solved by draining the toxic components in the soil. Preventive conservation is much less expensive than restorative conservation. More important, as has been said, "Some ecologies once destroyed by man can never be brought back no matter what we do." All our billions, all our technology can never bring back the tons of topsoil from the Gulf of Mexico to the American heartland. Nor can we ever bring back a single acre of wilderness once it is destroyed. There are no instant ecologies or instant forests. We must assess each new and old technological development for its ultimate impact on man.

* * * *

The Globe Is Circled with a Girdle of Filth

Louis J. Fuller

Mr. Fuller, why, are you so rough on the auto manufacturers?

FULLER: I can't clean up air pollution in Los Angeles County because the biggest portion, 90%, is from motor vehicles, over which I have no control. Stationary sources—industrial, commercial, residential—contribute only about 10% of our pollution.

Only 10% is from nonauto sources? That seems low.

FULLER: It is low. It's the result of 22 years of very aggressive action and the expenditure of millions of dollars. Industry in the Los Angeles basin is the most vigorously controlled of any I know. Our rules and regulations make it simply impossible for anyone to violate the law for any period of time. All large companies capable of producing air pollution are connected by radio transmitter to our dispatch center. Within seconds we can put into effect curtailment or shutdown plans developed jointly by the district and the companies. We've applied controls to the obvious—petroleum refineries, incinerators, chemical plants, open-hearth furnaces, auto-assembly plants—and to the not-so-obvious—restaurants, crematories, housing-tract developers who used to clear acreage with open fires.

What powers of enforcement do you now have?

FULLER: I can take criminal action against offenders, and have. Any inspector or engineer from the district observing a violation can issue a citation. We have a 97% conviction rate on cases we've brought

From *Forbes*, 104:55-57 (December 15, 1969).

to court. There are no favorite sons; we don't back off from anybody. Size makes no difference, nor importance.

What do you do? Slap their wrists? Or do you really come down on them?

FULLER: Six months in jail or $500 fine per day of violation, or both. But the most effective penalty is publicity. The newspapers used to print box scores on the front page, listing the companies fined.

So why don't you clean up auto exhaust pollution?

FULLER: That responsibility is given by state law to the State Air Resources Board, appointed by the Governor. I've raised more hell with the automobile industry than any other individual in the United States, but I have no authority over them. I *do* know that the 1966, 1967, 1968 and 1969 car models are violating state standards of exhaust emission.

What do the auto companies say when you tell them that?

FULLER: At first they denied it, but then they admitted it. In 1966 they said, "This is the first year, and like Avis, we'll try harder; wait till next year." The next year's models, and every one since, continued to violate the standards. I attended a meeting recently at which General Motors made the major presentation, and they admitted motor vehicles were responsible for most of the air pollution here.

What do auto companies say to the specific charge that their exhaust-control systems don't meet the state standards?

FULLER: They say, "Well, it's a difficult problem." And they say that one device can't control the two major polluting compounds, unburned hydrocarbons and oxides of nitrogen—in fact, that a device that takes care of one increases the other.

So why not two devices?

FULLER: Good question, and it was answered by Atlantic Richfield. They came up with a device that was 80% to 85% effective in controlling oxides of nitrogen, one smog ingredient. The company spent thousands of dollars developing it in the public interest. They took it to Detroit a couple of years ago, and that's the last we've seen of it. You attempt to find out information about it and the auto companies say, "Well, it doesn't quite do what we want it to."

What, specifically, would you do?

FULLER: The burden must be placed where it belongs, right back on the automakers. They should be required to test every vehicle they want to sell in the state. If they fail to meet our standards, somebody has to have the guts to say that they can't sell their cars here anymore.

The Justice Department has decided not to prosecute the automakers for conspiracy to delay development and installation of antismog devices. Does the Consent Decree ending the Government's case against the automakers reduce the chances of successful court action by other jurisdictions, such as New York City and Chicago?

FULLER: If the Antitrust Division of the Department of Justice of the Government of the United States was unable or reluctant to proceed, I don't see how some lesser agency of government would be any more successful.

Is the real answer an electric car? Or a steam car? Or a gas turbine car?

FULLER: That won't be necessary. The internal combustion engine, as it now exists, but using natural gas as a fuel—the same kind used in home stoves—would solve the air-pollution problem. The conversion would cost about $300 per car. And natural gas has advantages: It's cheaper, crankcase oil doesn't have to be changed so often because it isn't diluted, plugs would last much longer because they don't become corroded.

What about the suggestion that oil companies be required to produce less volatile gasoline that would reduce hydrocarbon emissions?

FULLER: Automakers often bring up that idea, but through years of testing we've proved that wouldn't help. Those hydrocarbons—butane and pentane—are so low in reactivity that they don't enter into our photochemical smog formation.

What happens, exactly, in the formation of smog?

FULLER: In effect, there's a chemical factory in the sky. Automobiles, burning gasoline, emit tons of hydrocarbons—that is, gasoline discharged unburned—and nitrogen oxides [created by high-temperature combustion]. There's a complex series of interactions between certain types of hydrocarbons, nitrogen oxides and ultraviolet radiation by the sun. The result is ozone, which causes eyes to water and throats to become scratchy and all the other familiar physical irritations of photochemical smog.

So if we didn't have so much sun, we wouldn't have so much smog?

FULLER: That's right; smog is not formed at night or on an overcast day. But we can't take action against the sun, so we're working on the other ingredients.

Well sure, this is rough, but can we depollute without wrecking the economy?

FULLER: If the medical profession decides we can't survive further air pollution, and something drastic has to be done—well, so be it. Doctors make such decisions all the time, in saying to a patient, "I'm sorry to tell you there will have to be an amputation." If there has to be an amputation of motor vehicles, so be it. Driving, after all, is a legislated privilege, not a right.

What does the medical profession say now about the effects of smog?

FULLER: Doctors are the first to admit they're not sure of the long-term medical effects of air pollution. We do know what happens when you're confined in a garage with the door closed and the engine running. What we don't know is what long-term exposures to lower levels will do. A carbon monoxide concentration of 30 or 40 parts per million parts of air over a sustained period of time may be equivalent to losing a pint of blood, because of a decrease in the hemoglobin's ability to transport oxygen. And the levels in car interiors on the freeways often reach 80, 90, 100 parts per million. There is evidence, of course, that air pollution is responsible for the increase in upper respiratory diseases. The Los Angeles County Medical Association decided that smog may seriously affect the lungs of young people. In fact, children through high school aren't allowed to exercise at all when the ozone level reaches 0.35 parts per million.

Los Angeles is making a strong bid for the 1976 Olympics. Does it make sense to bring the world's greatest athletes to a city where it's sometimes dangerous to breathe?

FULLER: Unless you put the games in the middle of the Gobi Desert, you'd have a hard time finding an area anywhere without smog. Of course it would be a risk; under adverse meteorological conditions we might get socked in here to a point requiring a postponement of some of the games.

Is the Gobi Desert the only place left in the world with clear air?

FULLER: The problem certainly is not confined to Los Angeles. The globe is circled with a girdle of filth that stretches from east to west, and as it travels it picks up additional millions of tons of pollutants. This creates what scientists call the greenhouse effect, a warming trend that could melt the polar ice caps and dangerously raise sea levels.

That sounds pretty grim.

FULLER: It's possible we may have proceeded so far now in the destruction of our environment that the environment's ability to heal itself is now less than man's ability to hurt it. Nature can be pretty rough when its balance is disturbed. We got a kickback once from nature—the Midwestern dust bowls. Of all the world's problems, I think air is the most critical, because a few minutes without proper air is fatal.

Clearly you don't think Detroit is moving fast enough on the problem. How about the rest of the world?

FULLER: The World Health Organization is making some attempts, but the solution has to come from the superpowers. I've talked to scientists from the Soviet Union and many other countries, and we're all in agreement about what's happening—but there's no way to stop it. There must be an international congress for the protection of the environment, and it must be more than just another organization with a name and no power. There is great alarm and concern all over the country, all over the world, on this subject, and the public is looking for leadership from somewhere. And it can't come from words alone.

Green Light for the Smogless Car

John Lear

Perhaps because they don't read the *Steambus Newsletter,* most newspaper editors across the country haven't yet told their subscribers that a green light is beckoning faintly through the smog in California to those rare motor vehicles that properly can be called polite: the ones that go about quietly, don't reek of body odors, and don't belch repulsive gases into people's faces.

Because I am a reader of the *Steambus Newsletter* and have some acquaintance with the purposes and persistence of its publisher and editor, I am not only aware of the appearance of the "go" signal but have been persuaded that the fume-free automobile the *Newsletter* is concerned with will within two decades at most—possibly within one—completely supplant the gasoline-burning buggy made popular by the first Henry Ford more than half a century ago.

There is no suggestion here that owners of common stock shares in conventional motor-making corporations such as Ford, General Motors, and Chrysler have cause to fear precipitate decline in the values of their holdings. On the contrary, I find growing cause to assume that present-day motor makers will evolve new designs to accommodate the evolving demand for clean breathing space just as fast as the demand asserts itself in the practical terms of the market place.

Private profit-seeking enterprise must by nature pursue research directions that promise reasonable return for money spent, and expressions of willingness to pay premium prices for smogless motoring were unheard from buyers until quite recently. Now, however, one

From *Saturday Review,* 52:81--86 (December 6, 1969).

premium offer is on record from the California State Legislative Assembly, and another is under consideration by the United States Congress. When these realities become widely recognized, popular resistance to slow strangulation by air pollutants will certainly expand first the public and then the private market for vehicles so built as not to create smog. To believe otherwise is to count the American citizen stupid and suicidal.

The editorial policy of the *Steambus Newsletter* holds that the promise of fume-free propulsion is sufficiently compelling in itself to stimulate widespread participation in smog prevention if word of new advances in transportation technology is spread as the improvements appear.

. . . [O]nly a limited time remains to prevent major American cities from being abandoned as centers of civilized living. California does not lightly dismiss the fact that smog contributes to the frequency of lung cancer, emphysema, chronic bronchitis, and asthma, and that in one year physicians advised 10,000 of their patients to move away from Los Angeles to protect themselves from the consequences of smog. The emphasis of *Newsletter* items, therefore, is on the speed with which the situation can be cleaned up. To obtain general cognition of progress, the first step is to equip a forty-four-passenger bus, operating on a regular schedule as part of an urban mass transit line, with an engine that emits virtually no fumes. To gain time, only the engine will be changed; the rest of the bus will be like all the other buses on the line. To gain still more time, the engine will be made of off-the-shelf parts already proved dependable in hundreds of thousands of miles of travel under normal traffic conditions on the open road and in congested city streets.

A certain amount of sophistication is expected of *Steambus Newsletter* readers. They are assumed to know that the only engine that fits these requirements is an *external* combustion engine, and they are also assumed to understand how an external combustion engine differs in principle from the *internal* combustion engine used in the standard gasoline auto of today. Since the distinctions are multiple, it may be well to review them here.

Internal combustion means just what the words say: The burning of fuel takes place *inside* the engine. Gasoline is fed from the fuel tank into the carburetor when the car driver steps on the accelerator. In the carburetor, the gasoline is mixed with air to facilitate subsequent burning. The vapors are then fired by electrically generated sparks from the spark plugs, and the resultant explosion drives pistons to turn a geared set of rods (the transmission) that rotates the axle and moves the car.

Not all of the energy released by the explosion of the gas is spent in driving the pistons; the excess goes out the exhaust pipe along with microscopic debris of the explosion.

As a piece of thermodynamic technology, the internal combustion engine is not essentially efficient. A variable fraction of its output is wasted in passing through the transmission. In slow traffic the fraction is large; at high speeds the fraction falls to perhaps 10 per cent. Consistently even burning of the gasoline is achieved only by the addition of lead to the fuel. But sixty years of human ingenuity have made the engine work incredibly well in spite of its shortcomings, much as nature through eons of time has enabled the bumblebee to carry on the miracle of pollination even though the laws of aerodynamics deny the probability of bumblebee flight.

The external combustion engine, as its name specifies, runs on fuel that burns *outside* the engine. Any one of a variety of inflammable liquids can be stored in the fuel tank and fed to a single spark plug. After the liquid is ignited, the flame burns evenly in the open, just as the pilot light does in a household oil burner. There is no need to add lead to the fuel, hence no lead fragments to be freed into the air. Because there is no explosion, there is no other debris to be disposed of through the exhaust pipe.

The simplest type of external combustion engine is the steam engine. Water is kept under pressure in a generator that resembles a stack of pancakes, each pancake consisting of a concentric coil of metal tubing. The water flows through the tubing toward the flame, warming the while, until evaporation occurs. When the car driver steps on the accelerator, a valve turns, shooting the steam into the engine pistons within thirty seconds after ignition. As the pistons move, the steam expands, losing force and allowing the pistons to push the now condensing vapor back into another reservoir, whence the water returns to the generator for recycling.

Like the internal combustion engine, the external combustion engine puts more energy into a piston stroke than is required to complete the stroke. However, the excess energy of the external combustion engine is not wasted through the exhaust pipe, but is returned to the system as heat in the water. Therefore, the external combustion engine must include a condenser coil large enough to dissipate the heat quickly. In the present state of steam engine art, most steam engines are bulkier than an internal combustion engine of equivalent power. So far as inherent efficiency is concerned, the external combustion engine is superior because it needs no elaborate transmission system: The engine's power can be applied directly to the turning of the axle. Braking is also more effective, because reverse power can be applied to the wheels as quickly as forward drive.

Steambus Newsletter readers not only are assumed to know about the advantages just described. They are still further assumed to know the early history of automobiling, in which steam-powered vehicles preceded gasoline-powered cars but fell behind in development because the external combustion engine car makers were not interested in mass production and so could not compete in price with Henry Ford once his assembly line got rolling.

Since the *Newsletter* has more readers within California than outside the state, most people on the subscription list know without being told why California has a special concern about smog. Although smog appears almost any place where population thickens, it persists more tenaciously in geographic bowls where prevailing winds trap the air that lies close to the earth's surface. Meteorologists call these traps atmospheric inversions because cool air, which normally falls, overlies warm air, which normally rises. The longer such a trap remains closed, the fouler the air caught in it becomes. The most densely peopled such bowl in the United States is the one containing the sprawl of Los Angeles.

At least as early as the year 1946, the *Los Angeles Times* was publishing reports beginning with such sentences as these: "Like a dirty gray blanket floating across the sky, a dense eye-stinging layer of smog dimmed the sun here yesterday. The fumes hung unmoving in the still air, raising tears and sniffles in thousands of Angelenos." As early as 1947, the city of Los Angeles organized itself as an air pollution control district—three years before a similar move was made by neighboring Orange county, five years before San Diego and San Francisco.

In those days, the highly unscientific consensus was that smog came principally from factory smokestacks. Factories were convenient scapegoats. They couldn't talk back. Critics found it easy to get rid of the problem by assailing the factory corporations, which had no faces but lots of money. This Alice-in-Wonderland approach to smog ran head on into a curious Dutchman named Arie Jan Haagen-Smit, who migrated from the University of Utrecht to the California Institute of Technology in pursuit of an understanding of hormones that regulate growth of living green plants. He wanted to learn the answers to such questions as why an onion smells different from a pineapple. He was distilling three tons of Hawaiian pineapples into a few grams of the essential oils that produce the flavor and the smell of that tropical fruit when his nose became offended by the smog he encountered en route to his laboratory.

What he smelled that day in 1948 obviously was neither pineapples nor onions. It seemed more like chlorine than anything else he had ever dealt with. What could it be? He opened a laboratory window that faced a Pasadena street and rigged up his instruments in such a manner as to draw into them a mass of air equivalent to the amount a person breathes in a day. Out of this stinking cargo he collected a few ounces of condensed smog. Analysis showed the condensate to be mostly ordinary water containing a number of foul-smelling chemicals—aldehydes, acids, and organic peroxides. All these substances had been known as products of incomplete burning and had been identified as causes of eye irritation. But none of them had ever been associated with air pollution.

The implications would have been obvious even to an insensitive laboratory hack. Haagen-Smit stood far above the hack level. Besides, he loved flowers and grass and trees, which were being blighted by smog. A vigorous conscience long before had carried him into the wildlife conservation crusades of the Sierra Club. The stuff he brought into his window jarred him off the faculty of Cal-Tech for a year to study polluted air.

Haagen-Smit returned from that working sabbatical convinced that smog is manufactured by sunlight operating on oxides of nitrogen to convert hydrocarbon fragments of organic matter that have been released into the air. His findings, which for the first time tagged automobile exhaust as a major source of smog, were disputed when he originally published them in scientific journals. Within two or three years, however, others confirmed his experiments.

Out of Haagen-Smit's research came gradual recognition that the motorcar produces 60 per cent of the air pollution in this country—in urban centers as much as 85 per cent. The ninety million tons of waste that pours into the atmosphere through auto exhaust pipes each year is triple the amount from any other source. Factory smokestacks annually belch thirty million tons, power plants fifteen million tons, space-heating furnaces eight million tons, and refuse incinerators three million tons.

With more motor vehicles under registration than any other county in the country (the 1966 figures were 2,932,980 cars and 436,218 trucks), Los Angeles County needed no shouting of the news from Haagen-Smit's lab to waken local public officials. In 1953, they appealed to automobile makers to undertake voluntary application of Haagen-Smit's discovery. The car makers pleaded surprise. In 1955, automobile exhaust fume-control bills began popping up in the California state legislature. In 1959, the first emission-control law was passed. By 1965, Los Angeles County, having lost patience with Detroit's failure to provide meaningful relief in a dozen years' time, took to the United States Justice Department a formal request for a grand jury investigation of the situation. Indictments on charges of criminal conspiracy were sought. But after two and a half years of hearings the Justice Department instead filed a civil suit in Central California District Federal Court at Los Angeles in January 1969. All the major auto makers were accused of ganging up to frustrate general use of fume-control devices on automobiles. Violation of the Sherman Antitrust Act was alleged.

Detroit attorneys argued that what the auto makers were charged with did not constitute restraint of trade, but on behalf of their clients they offered to sign a consent decree promising that no such collusion would be engaged in. The Justice Department accepted the offer. Although the decree was objected to by Los Angeles County, the State of California, and local governments as far away as the Pennsylvania towns of Erie and Lancaster, the document was signed by Federal District Judge Jesse W. Curtis during the last week of October. In an oral opinion, Judge Curtis ruled that the consent decree would not prejudice treble damage claims the concerned governments declared an intent to enter. The judge said that all the data that had been presented to the grand jury were open to subpoena in support of such claims, and that even the grand jury transcript would be available to those who could demonstrate a "need to know" the gist of the secret testimony.

Whatever comes of the talk of treble damage claims, the motor makers henceforth will not be in a favorable position to resist public demands for a sharp reduction in auto fumes. Engineers in Detroit are already scrambling to meet the limits fixed by the California legislature and enforced by the Air Resources Board chaired by Cal-Tech Professor Haagen-Smit.

Having spent sixty years and billions of dollars to make the gasoline-fueled internal combustion engine a technological work of art, the car makers are understandably loath to abandon their favorite offspring and begin lavishing attention instead on another creature of more stolid though healthier stock. They insist that they can reduce the gasoline engine's effluents to tolerable levels simply by adding devices that will enable the burning of progressively greater fractions of the gasoline.

Haagen-Smit, who once was characterized by a friend as a cross between Old Dutch Cleanser and St. George, sees an ultimate weakness in this approach. To him, there is no escape from the genetics of combustion. All burning depends on the presence of oxygen. As more oxygen is introduced to complete the destruction of hydrocarbons that now escape from the motorcar, the oxygen combines not only with the hydrocarbons but with nitrogen in the air, creating oxides of nitrogen that are among the most vicious elements in smog.

Even if the car makers manage to do everything they hope to be able to do, no way has yet been found to guarantee that, as time passes and a car's mileage mounts, the fume-control devices on it will not become less effective, inevitably allowing more than the legal amounts of fumes to escape. Nor is there any way to prevent unscrupulous drivers from removing the devices in order to get more power from the engine. There is also the problem of controlling emissions from cars registered outside California, where restrictions on fume emissions are less stringent. Finally, granting perfect performance, control devices will never be able to reduce pollution far enough to outbalance the expected doubling in number of cars on the roads by the year 1980. Most cars operating in any year are not new cars, hence are not yet equipped with the most effective devices. Haagen-Smit estimates that by 1980 enforcement of the laws now on California's books will eliminate about 60 per cent of the waste emitted by motorcars. "But control will always be marginal," he told a report for Cal-Tech's *Engineering and Science*. "It is always going to be an uphill fight."

When Haagen-Smit says "fight" he intends a vigorous definition of the word. One automobile maker, who thought otherwise and neglected to comply precisely with Air Resources Board standards, found the California market closed to its newest models during the period of weeks it took to correct the shortcomings. Yet, for all his reputation as a scouring agent, Haagen-Smit is a pillar of patience when measured alongside some other California opposers of smog. State Senator Nicholas Petris of Alameda has vainly sponsored a law that would force internal combustion engines from California roads after 1975. State Assembly Transportation Committee chairman John Francis Foran, less extreme but more persistent, has conducted public hearings on the advantages of external combustion engines. He is the father of three pieces of smog-fighting legislation: one fixing finite limits on fume emission by gasoline cars (with standards twice as tough as those later adopted by the federal government), another encouraging rapid development of all types of low-emission vehicles by authorizing the State of California to pay as much as a 100 per cent premium for such cars in up to 25 per cent of all purchases for the state-owned automotive fleet, and a third requesting the California Highway Patrol to test six patrol cruisers equipped with steam engines at the expense of General Motors. When General Motors, for undisclosed reasons, withdrew from the last-named project, Foran took the unusual step of appealing directly to Washington for an experimental grant from the Department of Transportation. No state legislature had ever before been given federal money. That Foran got some for his committee is an index of the extremity of the smog dilemma.

The contract went to a welding of mid-twentieth-century nuclear science with the long vanished art of custom building steam-powered limousines. Perhaps the only place in the country qualified for such an assignment is the office of William M. Brobeck and Associates, of Berkeley. Just after being graduated from Stanford University and M.I.T., Mr. Brobeck worked in one of the last steamcar shops run by the late Abner Doble and his brothers. Others had originated the idea of replacing the horse-drawn buggy with steam-propelled carriages. Literature on the subject mentions a French steamcar in 1770, a Gurney Coach in 1827, an Austin in 1863, a Gentleman's Speedy Roadster in 1896, a Locomobile in 1898, a White in 1904, a Clarkson Bus in 1904, and many others. The first "big" steamcar builders were the Stanley twins, Freelan and Francis, both inventors. They designed a radically new little buggy that got around fast but depended for mileage on frequent refillings of the steam boiler at public troughs provided in those days for horses. The Stanleys were happy to turn out 650 cars a year, the number that moved off the Ford assembly line every day. The Stanleys made comfortable profits on their timetable and had no wish for more. Consequently, when public water troughs were removed from most towns because of an epidemic of hoof-and-mouth disease in the early 1900s and the cities of Boston and Chicago about the same time closed their streets to the steamers because (lacking condensers) the cars billowed such dense clouds of exhaust steam in cold or damp weather that visibility of following vehicles was hampered, there was no popular demand to keep the Stanleys out of receivership in 1923. The Dobles had a more modern view. Their Model E Simplex was the first sustained attempt that anyone had made to bring the price of steam motoring within mass market reach.

The Doble Simplex, priced at $2,000, never came out. But the Model E Deluxe is still the most talked about of all steamcars, with the possible exception of the Stanley Rocket, which was clocked at 127.66 miles per hour in 1906, was estimated to have gone 150 mph, and was said once to have gone so fast that it left the ground briefly.

By the 1930s, when Brobeck got into the picture, the internal combustion engine was proving so popular on the highways that it was beginning to invade the railroads, where the steam locomotive had always been supreme. Diesel oil was being introduced as a locomotive fuel. To fight off the diesels, Abner Doble's brother Warren built a 500-horsepower steam engine to drive a two-car passenger train built by Budd in Philadelphia. The New Haven Railroad ran the train experimentally between Waterbury and Bridgeport, Connecticut. Brobeck was responsible for a major part of the design of this engine, the laboratory testing of its boiler and auxiliaries, and direct technical management of the installation and maintenance of the power plant during regularly scheduled railroad service. When the twin car failed to achieve its purpose, the young man went west from Connecticut, driving a Doble steam-powered bus from Davenport, Iowa, to the last Doble factory at Emeryville, California, where the Besler Company, taking over what was left of the Doble enterprise, dismantled the bus.

With four years in steam propulsion behind him, Brobeck was caught up in the exploration of a radically new power source: the nucleus of the atom. He helped Ernest O. Lawrence build the cyclotron and was himself chiefly responsible for designing the bevatron. After the Lawrence lab became the University of California Radiation Laboratory, Brobeck advanced to associate director and chief engineer of the institution. In 1957, he left the lab to set up the consulting firm that now bears his name.

About two years ago, when particle accelerator research reached a plateau, Brobeck and his associates surveyed likely opportunities to diversify their consultations. The steam-powered automobile was a conversation piece at the time. The central problems of steam propulsion lay in the very areas that had been most exhaustively analyzed by nuclear scientists—technological manipulation of the laws of thermodynamics, effective transfer of large amounts of heat, strengthening of construction materials, development of metallic alloys resistant to corrosion, complex mechanical and electrical and electronic controls, miniaturization of components, simplified packaging of delicately tuned devices. [A] call for bids on the steambus seemed an ideal target, and the Brobeck firm accelerated itself at the precise angle and speed to score a hit.

Thirty years after seeing the steam engine he built for the Dobles supplanted by diesel power, Brobeck now will reverse the roles and put a steam engine into a bus to replace a diesel. Only the engine will be changed. The transmission will remain intact so that the bus drivers will not be required to learn the markedly different pacing that would be occasioned by its removal. To save weight, the fuel tank capacity of the bus will be halved. To dispose of the excess heat, a steam condenser three times as big as the radiator presently on the bus will be installed.

Brobeck is confident that steam power will work satisfactorily on buses and trucks even in the relatively primitive state of steam technology today. The private passenger car, however, is a different breed of vehicle, and Brobeck confesses to a hearty suspicion that car-buyer response to steam power will depend on extensive technological refinement. After all, only twenty years of work (a third of the time spent on the gasoline auto) is represented in the Doble Model E—a copy of which, incidentally, Brobeck sends skeptical visitors to see at the home of Bernard Becker in Walnut Creek near San Francisco. The points at which new research ought to be directed will become evident only after prototype steamcars are put on the open road and into urban traffic snarls. California Assemblyman Foran thinks that highway patrol cars would give steam engines about as rugged a tryout as could be imagined. According to the terms of a resolution he argued through the Assembly some months ago, the California Highway Patrol will test two steam-powered cars for six months and report the performance to the legislature. Favorable reports would be priceless advertising for the engine designers, to whom the cars will be returned after the tests.

* * * *

In five months of gathering information about external versus internal combustion engines, I have met with, talked to, or heard of no one who claimed that an economical steamcar could be rolled off an assembly line onto the road in less than five years. I have heard claims that it could be done in ten years. The best informed independent advice tends to clump around a fifteen-to-twenty-year interval. I also have heard arguments purporting to show that steam will never do it but that vapors heavier than water vapor surely will.

* * * *

The editor and publisher of *Steambus Newsletter* will not be at all chagrined if an alternate to water vapor turns out to be the most efficient fluid for an external combustion engine. To them, the important aspect of their experiment is its effect on smog. To emphasize this point, Assemblyman Foran's Transportation Committee scheduled public hearings on December 4 in Los Angeles on the dangers of lead poisoning being spread through the atmosphere. The intent is to discourage use of lead in gasoline, and so to put the internal combustion engine on a fairer competitive footing with external combustion engines, which do not require leaded fuel.

On the principle that anything learned from steam propulsion will inevitably advance the understanding of external combustion engines generally, Foran intends to go back to Washington to ask that his $750,000 steambus grant be doubled to cover three additional experimental contracts that have been approved by his expert advisory panel on condition that money be available. First in line for these contracts is Steam Power Systems, Inc., at San Diego. This firm's expertise is centered in the person of Ken Wallis, who left William Lear a year ago after directing the development of a unique triangle-shaped steam engine that Lear had hoped to put into a racing car on the Indianapolis Speedway last Memorial Day. The engine, which Lear called the Delta, worked beautifully until someone ran it to impress visitors and forgot to open the oil injector. Since Lear owns the Delta design, Wallis cannot copy it without infringing patents, but he can apply his knowledge of steam locomotion to bus engines.

One of Wallis's competitors in the California steambus experiments is Doug Paxton, an inventor whose engine runs on a commonly known chemical vapor and is operating in the shop of Paxve, Inc., at Newport Beach, California. Another entry in the competition is Steam Engine Systems, of Newton, Massachusetts, a corporate brainchild of M.I.T.'s Professor Morse, who has no hardware to show yet, but who commands a brilliant constellation of laboratory skill.

The California initiative to clear the air around us will be felt increasingly throughout the country in the months ahead. Residents of many states will see fume-free cars in operation and will be able to judge for themselves the appeal of quiet, odorless motoring. The examples will be provided by local service fleets of gas and electric companies. These cars will use a natural gas fuel system initiated by Pacific Lighting and Service in Southern California. The natural gas is stored in an auxiliary tank in the car separate from and in addition to the conventional gasoline tank. A flick of a switch transfers the fuel feed line from the gasoline tank to the auxiliary and back again. Natural gas is already used to carry visitors around Disneyland and is being tested on a small number of federal government vehicles by the General Services Administration.

* * * *

A Vapor Moving North-Northwest

Daniel Lang

A few moments after the underground nuclear blast known as Project Gnome went off, at noon on a Sunday, in December, 1961, in a flat and chilly stretch of desert southeast of Carlsbad, New Mexico, all of us who were watching the event from a mound of bulldozed earth four and a half miles due south of ground zero—some four hundred foreign observers, congressmen, government scientists, local citizens, photographers, and reporters—could tell that something had gone wrong. What gave us this impression was not the broad blanket of dust that the explosive—deep below in a formation of salt rock—had jolted out of the desert. Nor was it the bouncing we took—the result of a violent earth tremor that had been caused by the nuclear charge, which was one-fourth as powerful as the Hiroshima bomb. (In the immediate vicinity of the explosion, the desert leaped three feet, and it has yet to descend to its former level.) We had been told to expect these things. Rather, it was the sight of thick and steadily thickening white vapor at the scene of the firing that made us think that plans had miscarried. The vapor was puffing up through an elevator shaft that dropped twelve hundred feet to an eleven-hundred-foot tunnel, at the end of which the explosive, and also much of the project's experimental equipment, had been installed. As we watched the vapor slowly begin to spread, like ground fog, and, rising, vanish into the air, we knew we were witnessing something that we had been practically assured wouldn't happen—venting, or the accidental escape of radioactivity into the atmosphere. "The probability of the experiment venting is so low as to approach the im-

From *An Inquiry into Enoughness*, by Daniel Lang (New York, 1965), pp. 1-15.

possible," the Atomic Energy Commission had stated in a comprehensive pamphlet it had published on Project Gnome. Indeed, at a briefing held the previous evening in Carlsbad, where Gnome's headquarters were located, one of the speakers had warned that the shot was just a small one and might well disappoint us as a spectacle. It was the excitement of its underlying idea that made it worthwhile for us to be at the proving ground, we had been told, for Project Gnome marked the opening of the Plowshare Program—a series of nuclear blasts whose purpose, as the name implied, was to turn the atom to peaceful ways. Any number of benefits, we were informed, could flow from these blasts: harbors might be carved out of wasteland in Alaska; oil might be dislodged from shale; abundant sources of water under great mountains might be freed; diamonds might be made out of ordinary carbon.

We were in no danger—the wind was blowing the vapor to the north-northwest of us—but the feeling seemed to take hold that this wasn't necessarily the Prophet Isaiah's day. Before the explosion, a gala mood had prevailed on our barren mound. Local ranchers, their big Stetsons bobbing, had heartily declared that it was a great day for these parts. The operators of nearby potash mines—the world's largest producers of this chemical—had agreed. Their wives, modishly clad, had greeted each other effusively. And Louis M. Whitlock, the manager of the Carlsbad Chamber of Commerce, had assured me, "This bomb is for the good of mankind, and we're for it," as we awaited the explosion. Representative Ben Franklin Jensen, of Iowa, a Republican member of the House Appropriations Committee, had also caught the proper spirit. "There are certain things you just have to spend money on, and Plowshare is one of them," he told me. The foreign visitors lent a certain glamour to the occasion. There was Professor Francis Perrin, for instance—a small, goateed man with elegant manners who was the High Commissioner of France's Commissariat á l'Energie Atomique. The science attaché of the Japanese Embassy was there, too—a young chemist named Dr. Seiichi Ishizaka. Chatting with him shortly before the venting, I had gathered that his government was of two minds about the wisdom of the day's explosion. "Japan is curious," he had told me, smiling politely. The bustle of the many journalists on the scene had added to the festive air. The local people had been fascinated by their activities, clustering around each time Dr. Edward Teller, the widely celebrated father of the H-bomb, who is also the father of Plowshare, posed for television crews. On the high-school platform in Carlsbad during the previous evening's briefing, he had, in response to a reporter's question, agreed that the Plowshare Program was "too little and too late," and referring to the recent resumption of atmospheric testing in the Soviet Union, had gone on to say, "Plowshare had to wait for permission from the Kremlin, which it is giving in a slightly ungracious manner."

Now, as the insidious gases continued to escape from the shaft, the gala mood faded. An A.E.C. official, speaking over a public-address system from a crudely constructed lectern, announced that all drivers should turn their cars around to facilitate a speedy retreat from the test area. An evacuation, he said, might be in order. A short while later—about half an hour after the detonation—the same official, a calm, affable man by the name of Richard G. Elliott, announced that, according to word from a control point a hundred yards forward, the venting had created a radioactive cloud, low and invisible, which was moving in the general direction of Carlsbad, twenty-three miles away to the northwest. The invisible cloud, which was being tracked by an Air Force helicopter equipped with radiation counters, was expected to miss the town, but it would pass over a section of the highway on which we had driven from Carlsbad. The state police had consequently been instructed to throw up a roadblock there. Until futher notice, the only way to reach Carlsbad would be to head southeast and follow a detour of a hundred and fifty miles. Some spectators left at once to take this roundabout route, figuring that they might as well get the trip over and done with, rather than face an indefinite delay. Some other spectators also departed hurriedly; they suspected the A.E.C. of being excessively cautious, and hoped to use the direct highway to Carlsbad before the police could organize their blockade. As things turned out, a few of these motorists did elude the police, only to be intercepted eventually in Carlsbad itself. Seven cars were found to be contaminated; the A.E.C. paid to have them washed down. Two of the passengers, according to the A.E.C., showed slight, easily removable traces of radioactivity, one on his hand and the other on his clothing and hair. As for the cloud, the helicopter that had started tracking it had been forced to return to base when the craft's instruments showed that it was being contaminated. Another machine took its place, and the pilot of this kept the cloud under surveillance until darkness forced him to give up his mission; the cloud was then five miles north of a small town called Artesia, about sixty miles north-northwest of the test site; it had hovered briefly over the eastern edge of the town, and continued in its north-northwesterly path. At the time he took his leave of the cloud, the pilot reported, its radiation was diminishing steadily—a process attributed to nature, rather than to Gnome's artificers.

Fortunately, the countryside over which this gaseous debris was being wafted was only sparsely populated. In fact, this was one of the reasons the explosive had been set off in this particular area. In spite of the reassurances about venting in the pamphlet, the A.E.C. and its chief contractor for Plowshare—the University of California's Lawrence Radiation Laboratory, in Livermore, California—had had this eventuality very much in mind when they planned Gnome. Many precautions had been taken. The tunnel was packed with bags of salt and blocks of concrete, designed to arrest the spread of radioactivity. Wind patterns had been analyzed by the United States Weather Bureau during the entire week before the shot. The day's detonation had, in fact, been delayed four hours until the winds were considered to be blowing in a safe direction. Ranchers for five miles around had been evacuated, tactfully, by being asked to join the Gnome spectators; their cattle, less privileged, had simply been driven off to roam different pastures for the day—or for however long it might take the United States Public Health Service to certify the cleanliness of their familiar acres. The Federal Aviation Agency had been asked to order planes in the area to maintain a certain altitude until further notice. The dryness of the salt formation notwithstanding, the United States Geological Survey had made ground-water surveys of the surrounding area for six months before the shot and would continue to do so for at least a year afterward, in order to keep tabs on any underground movement of radioactive material. Seismic effects had also been anticipated. A special bill had been put through Congress to assure the potash industry of suitable indemnification in the event of damage. On the day of the detonation, no potash miners were on hand to chip at the rose-colored walls of their rough corridors. Nor were tourists permitted to explore the Carlsbad Caverns, thirty-four miles to the east of the detonation site. Acting on behalf of Project Gnome, the Coast and Geodetic Survey had placed a seismograph inside the Caverns. A member of the Caverns' staff—a naturalist from the National Park Service—was on hand to measure seismic effects in his own way; he watched to see if the blast would ripple one of the still, subterranean ponds that had been created over millennia, partly by drops of water from the cave's stalactites. (It didn't.) In retrospect, perhaps the most significant of all the precautions taken was the relatively last-minute reduction of the yield of the explosive from ten kilotons, as originally planned, to five kilotons. "Whoever made *that* decision, I'd like to shake his hand," an A.E.C. official told me the day after the shot.

Those of us who, like me, were waiting for the roadblock to be lifted, passed the time as best we could. We discussed our reactions to the blast for a while, but, oddly, this soon began to pall. Some of us wandered over to a chuck wagon that the A.E.C. had thoughtfully laid on, and bought ourselves coffee and sandwiches. Now and then, we heard new announcements, of varying interest, on the public-address system. One dealt with the far-flung network of seismic recording stations that had been organized by the Department of Defense. A colonel mounted the lectern to tell us that the network appeared to have functioned well. (He didn't know then that Gnome's seismic signal had been recorded in Scandinavia and Japan.) The firing, the colonel added, had taken place "at exactly one four-thousandth of a second after noon." Returning to the lectern, Elliott told us that, according to the instruments, the radiation at the bottom of the shaft now came to a million roentgens an hour, while on the ground at the top of the shaft the count was ten thousand roentgens an hour—twelve and a half times the lethal exposure for a healthy man.

After a while, some of us went and sat in our cars to read or doze or just get out of the cold. Those who didn't could stare at the shaft, from which vapor was still issuing, or, if they preferred, scan the desert, stubbled with tumbleweed and greasewood and cactus. Only the distant sight of a potash refinery relieved the terrain. Bluish-white smoke was pouring from its tall chimney, its furnace having been left unbanked on this day of days. The refinery lay due northwest, near the Carlsbad road, so I knew that the radioactive gases were bound to mingle with the vapors of the tall chimney. Like my fellow-spectators, though, I had no idea when that would come to pass.

The technical objectives of the day's blast, which were almost entirely in the hands of Livermore scientists, were well planned, it had been impressed on all of us in the course of the briefings before the shot. The central purpose was to see what happened when an atomic explosive was set off in a salt formation—what is called phenomenology. The Livermore people hadn't previously had a chance for such a test, their underground efforts thus far having been limited to military shots in the volcanic tuff of the Nevada test site—a substance that doesn't retain heat nearly as well as salt does. And heat was the key to much of what the researchers were seeking to learn. Gnome would enable them to carry out a heat-extraction experiment, for example—the general idea being to investigate the possibility of tapping for productive uses the inferno of superheated steam and other forms of energy that would result from the detonation. This energy, it was hoped, would be contained in a cavity in the salt that the explosive, low though its yield was, would create in about a tenth of a second. The cavity, if it didn't collapse, would be egg-shaped and glowing, and it would be about a hundred and ten feet in diameter; six thousand tons of molten salt were expected to run down its sides and compose a pool thirty-five feet deep. The cavity would also be "mined," by remote control, for radioactive isotopes—unstable atoms that are produced by a nuclear explosion, a fair percentage of which are valuable in scientific research, medical treatment, and industrial processes. (One of them, strontium 90, which is greatly feared in fallout, may some day be used in long-lived batteries to power unmanned weather stations in god-forsaken regions, a Livermore expert told me.)

For pure researchers, it was thought, Gnome's most interesting data might be gained from the large numbers of neutrons—uncharged particles that are part of the atomic nucleus—that would be produced by the blast. In the instant of the explosion, I had been told, Gnome would release as many neutrons as a laboratory apparatus could release in several thousand years. So plentiful would they be, in fact, that only one out of ten million could be studied. Even so, much new light might be shed on such matters as the different velocities of neutrons and the interaction of these particles, which are usually emitted in bursts that last less than a hundred-millionth of a second, an interval of time that is known in scientific shoptalk as "a shake."

But these technical objectives of Project Gnome were only a part of the Plowshare Program, and the Plowshare Program was something more than a scientific enterprise—a fact that had become apparent in the days immediately preceding the desert shot, when Carlsbad had been rife with briefings, interviews, and informative handouts. The case for Plowshare, in the opinion of some of the foreign observers and other people I talked with, seemed to rest on a variety of grounds. I learned, for example, that the proposed series of blasts had been approved by the A.E.C. four years before, which raised the question of why they were being started at this particular time. Plowshare officials readily acknowledged that the complete answer certainly included the state of international affairs. Was Plowshare, then, a solid program or a passing, virtuous response to the Russian resumption of atmospheric testing? Perhaps Plowshare's name was partly to blame for this questioning attitude. "It sounds a little too much like magic," a foreign scientist remarked. "So many swords are being made just now."

In any event, a day or two before the shot, I discussed Plowshare in Carlsbad with two of its overseers, both of whom were strongly in favor of the program, as one would expect, but in a fairly thoughtful, unmagical way. One of them was John S. Kelly, a bespectacled, mild-mannered man of thirty-nine who directed the A.E.C.'s Division of Peaceful Nuclear Explosives. He saw Plowshare's explosives as scientific and engineering tools. It excited him, he said, to contemplate the excavation jobs that might be performed in the future, like blasting lakes out of the wilderness and breaking up ore deposits that could be leached out. Plowshare represented a continuation of the whole history of explosives, Kelly said. Certainly explosives could be harmful, he conceded, but on the other hand gunpowder had done away with the feudal system and TNT had made possible the mining of fossil fuels.

"But can we afford to guess wrong with nuclear explosives?" I asked. "Don't they represent an ultimate kind of energy?" "Why not use them for our ultimate good?" Kelly replied. For an undertaking concerned with the peaceful uses of the atom, I remarked, Plowshare appeared to have its ambiguities. The fissionable material and the equipment for the Gnome explosive, I mentioned, had been taken from our armaments stockpile; the explosive was being concealed from the public gaze, the same as a weapon is; men in uniform had come to Carlsbad for the shot, and were participating actively in its preparation; and among those prominently involved were people from Livermore, which was noted primarily as a center of weapons design.

Kelly was quick to grant that the line between the peaceful and the military sides of the atom was fuzzy. It would be nice, he said, if the two functions could be neatly demarcated, for in that case the Plowshare Program, living up to its name more fully, could have postponed the blasts until war was an obsolete institution. But that wasn't the way things were, in Kelly's view. "We may have to take our peaceful uses when we can," he said.

The other official I talked with was Dr. Gary H. Higgins, the director of the Plowshare Division of the Lawrence Radiation Laboratory. Higgins was a soft-spoken chemist of thirty-four, whose desk in his Carlsbad office was adorned, when I saw it, with a small ceramic gnome he had bought in a department store. Like Kelly, he believed that nuclear explosives had a great peacetime future. "Within five to fifteen years, they'll be basic to our industrial economy," he told me. "They'll help us get at raw materials we need for our growing population. It may take us time to make use of them. After all, forest husbandry developed only when the nation was practically deforested." He was delighted that the United States was moving ahead with Plowshare, but not, he told me, because it relieved him of his weapons duties at Livermore. The two kinds of work, he felt, were not pure opposites; there was a difference between weapons and war, he said, just as there was between a police force and murder. But whether an idea like Plowshare or an arms race was to dominate our lives in the years ahead was another matter. It depended, Higgins thought, on whether mankind could eventually achieve an immense self-consciousness. "It would not cater to the oversimplified images that religion and ethics tend to give us," Higgins said. "It would enable us to recognize our weaknesses. We'd know our motives for acting the way we do, and what else is it that counts but intent, whether shots are called Plowshare or something else?"

It was almost four hours after the detonation when I left the bulldozed mound in the desert. The roadblock hadn't yet been lifted, but to a number of us that didn't matter. We were chafing to get away, although not for any sensible reason I heard expressed. Perhaps the others felt, as I did, a sense of rebellion and indignation at being trapped by a mysterious, invisible antagonist. In the distance, the refinery's tall chimney continued to surrender its thick plume of smoke, giving no sign, of course, whether there had yet been any mingling with the radioactive cloud. Absurdly, I felt like going to the refinery to find out. Around us, shadows were beginning to fall on the desert, making it seem more limitless than ever, and underscoring our marooned condition.

At any rate, when a rancher who was among the spectators mentioned to some of us that certain back roads might bring one out on the Carlsbad highway three or four miles beyond the police blockade, I was off at once, in a car with two other men—Ken Fujisaki, a young correspondent for a Tokyo newspaper, the *Sankei Shimbun*, and David Perlman, a reporter for the San Francisco *Chronicle*. The rancher, who himself was in no hurry to leave, had said he hadn't used those particular back roads in fifteen years, but at the time this remark had struck us as irrelevant. Our immediate goals were a windmill and a gas well—two landmarks that, the rancher had said, might soon guide us on our way to Carlsbad.

"How would you like to spend two weeks in a fallout shelter?" Perlman, who was driving, asked me as he impatiently started the car.

After a ten-minute drive over a bumpy, rutted road, we were at the gas well. We were also at a dead end. As we were looking at each other in puzzlement, we heard the honk of a car horn behind us, and discovered that we had been leaders of men. Nine other cars had followed us to the dead end; we had been too intent on our flight from safety to notice them. One of the vehicles was a small orange government truck, and another was a sports car—a dirty, white Triumph whose driver wore goggles. Some of us got out of our cars, conferred ignorantly, and decided to go back and follow a dirt road that had intersected the one we were on. This road also came to a dead end. Backtracking, we tried another, and then another. The fourth ran parallel to a ranch fence, on the other side of which were cattle and horses. Beyond the field they were in we could see the Carlsbad highway, only a couple of miles off. The fence seemed to run on endlessly, leading nowhere. Our caravan halted, and a few of us climbed a stile to seek advice at the ranch. We found a young Mexican hand, who obligingly corralled the animals, and opened a gate into a muddy, reddish road that crossed the field. In no time we were on the highway to Carlsbad. To get there, we had gone east, north, west, and northeast. Now we passed the potash refinery, its tall stack still smoking. I looked at it as long as I could. No police intercepted us. When we reached the Project Gnome office in Carlsbad, we learned that the roadblock had been called off fifteen minutes after our departure. Perlman asked that he be gone over with a radiation counter. He proved to be fine, which meant the rest of us were.

When I arrived at my motel, the manager phoned me. He was a transplanted Englishman with whom I had made friends. Since I was leaving the next day, I thought perhaps he was calling to say goodbye, but it was Project Gnome that was on his mind.

"I'm sick in bed, you see, so *I'm* quite all right, but it's the staff—" he began. A guest, he said, had told the cashier in the restaurant not to touch the money of anyone who had been to the test. The cashier had become hysterical. Then a policeman had come and collected two other members of the staff to have them "counted" at the Gnome office; the two had been spectators at the shot and had been among those who eluded the roadblock.

"There's no need for any concern, is there?" the manager asked me uneasily. "I mean, those men out there know what they're doing, don't they?"

I could hear him breathing at the other end of the phone, waiting for my answer.

"Of course they do," I said. "Of course everything's all right."

PART III

WATER

My Dirty Stream

Sailing down my dirty stream,
Still I love it and I'll keep the dream
That some day, though maybe not this year,
My Hudson River will once again run clear.

It starts high in the mountains of the north,
Crystal clear and icy trickles forth
With just a few floating wrappers of chewing gum
Dropped by some hikers to warn of things to come.

At Glens Falls, five thousand honest hands
Work at the Consolidated Paper Plant.
Five million gallons of waste a day,
Why should we do it any other way?

Down the valley one million toilet chains
Find my Hudson so convenient a place to drain,
And each little city says, "Who, me?
Do you think that sewage plants come free?"

From the record *God Bless the Grass*, by Pete Seeger, Columbia CL 2432.

Out in the ocean they say the water's clear,
But I live right at Beacon here,
Halfway between the mountains and the sea.
Tacking to and fro this thought returns to me.

Well, it's sailing down my dirty stream,
Still I love it and I'll dream
That some day, though maybe not this year,
My Hudson River and my country will run clear.

Pete Seeger

Walden Pond

Henry David Thoreau

The scenery of Walden is on a humble scale, and, though very beautiful, does not approach to grandeur, nor can it much concern one who has not long frequented it or lived by its shore; yet this pond is so remarkable for its depth and purity as to merit a particular description. It is a clear and deep green well, half a mile long and a mile and three-quarters in circumference, and contains about sixty-one and a half acres; a perennial spring in the midst of pine and oak woods, without any visible inlet or outlet except by the clouds and evaporation. The surrounding hills rise abruptly from the water to the height of forty to eighty feet, though on the southeast and east they attain to about one hundred and one hundred and fifty feet respectively, within a quarter and a third of a mile. They are exclusively woodland. All our Concord waters have two colors at least, one when viewed at a distance, and another, more proper, close at hand. The first depends more on the light, and follows the sky. In clear weather, in summer, they appear blue at a little distance, especially if agitated, and at a great distance all appear alike. In stormy weather they are sometimes of a dark slate color. The sea, however, is said to be blue one day and green another without any perceptible change in the atmosphere. I have seen our river, when, the landscape being covered with snow, both water and ice were almost as green as grass. Some consider blue "to be the color of pure water, whether liquid or solid." But looking directly down into our waters from a boat, they are seen to be of very different colors. Walden is blue at one time and green at another, even from the same point of view. Lying between

From *Walden: A Story of Life in the Woods*, by Henry David Thoreau (New York, 1902), pp. 197-218. Reproduced here under the editor's title.

the earth and the heavens, it partakes of the color of both. Viewed from a hill-top it reflects the color of the sky, but near at hand it is of a yellowish tint next the shore, where you can see the sand, then a light green, which gradually deepens to a uniform dark green in the body of the pond. In some lights, viewed even from a hill-top, it is of a vivid green next the shore. Some have referred this to the reflection of the verdure; but it is equally green there against the railroad sand-bank, and in the spring, before the leaves are expanded, and it may be simply the result of the prevailing blue mixed with the yellow of the sand. Such is the color of its iris. This is that portion, also, where in the spring, the ice being warmed by the heat of the sun reflected from the bottom, and also transmitted through the earth, melts first and forms a narrow canal about the still frozen middle. Like the rest of our waters, when much agitated, in clear weather, so that the surface of the waves may reflect the sky at the right angle, or because there is more light mixed with it, it appears at a little distance of a darker blue than the sky itself; and at such a time, being on its surface, and looking with divided vision, so as to see the reflection, I have discerned a matchless and indescribable light blue, such as watered or changeable silks and sword blades suggest, more cerulean than the sky itself, alternating with the original dark green on the opposite sides of the waves, which last appeared but muddy in comparison. It is a vitreous greenish blue, as I remember it, like those patches of the winter sky seen through cloud vistas in the west before sundown. Yet a single glass of its water held up to the light is as colorless as an equal quantity of air. It is well known that a large plate of glass will have a green tint, owing, as the makers say, to its "body," but a small piece of the same will be colorless. How large a body of Walden water would be required to reflect a green tint I have never proved. The water of our river is black or a very dark brown to one looking directly down on it, and, like that of most ponds, imparts to the body of one bathing in it a yellowish tinge; but this water is of such crystalline purity that the body of the bather appears of an alabaster whiteness, still more unnatural, which, as the limbs are magnified and distorted withal, produces a monstrous effect, making fit studies for a Michael Angelo.

The water is so transparent that the bottom can easily be discerned at the depth of twenty-five or thirty feet. Paddling over it, you may see many feet beneath the surface the schools of perch and shiners, perhaps only an inch long, yet the former easily distinguished by their transverse bars, and you think that they must be ascetic fish that find a subsistence there. Once, in the winter, many years ago, when I had been cutting holes through the ice in order to catch pickerel, as I stepped ashore I tossed my axe back on to the ice, but, as if some evil genius had directed it, it slid four or five rods directly into one of the holes, where the water was twenty-five feet deep. Out of curiosity, I lay down on the ice and looked through the hole, until I saw the axe a little on one side, standing on its head, with its helve erect and gently swaying to and fro with the pulse of the pond; and there it might have stood erect and swaying till in the course of time the handle rotted off, if I had not disturbed it. Making another hole directly over it with an ice chisel which I had, and cutting down the longest birch which I could find in the neighborhood with my knife, I made a slip-noose, which I attached to its end, and, letting it down carefully, passed it over the knob of the handle, and drew it by a line along the birch, and so pulled the axe out again.

The shore is composed of a belt of smooth rounded white stones, like paving stones, excepting one or two short sand beaches, and is so steep that in many places a single leap will carry you into water over your head; and were it not for its remarkable transparency, that would be the last to be seen of its bottom till it rose on the opposite side. Some think it is bottomless. It is nowhere muddy, and a casual observer would say that there were no weeds at all in it; and of noticeable plants, except in the little meadows recently overflowed, which do not properly belong to it, a closer scrutiny does not detect a flag nor a bulrush, nor even a lily, yellow or white, but only a few small heart-leaves and potamogetons, and perhaps a water-target or two; all which however a bather might not perceive; and these plants are clean and bright like the element they grow in. The stones extend a rod or two into the water, and then the bottom is pure sand, except in the deepest parts, where there is usually a little sediment, probably from the decay of the leaves which have been wafted on to it so many successive falls, and a bright green weed is brought up on anchors even in midwinter.

We have one other pond just like this, White Pond in Nine Acre Corner, about two and a half miles westerly; but, though I am acquainted with most of the ponds within a dozen miles of this centre, I do not know a third of this pure and well-like character. Successive nations perchance have drank at, admired, and fathomed it, and passed away, and still its water is green and pellucid as ever. Not an intermitting spring! Perhaps on that spring morning when Adam and Eve were driven out of Eden, Walden Pond was already in existence, and even then breaking up in a gentle spring rain accompanied with mists and a southerly wind, and covered with myriads of ducks and geese, which had not heard of the fall, when still such pure lakes sufficed them. Even then it had commenced to rise and fall, and had clarified its waters and colored them of the hue they now wear, and obtained a patent of heaven to be the only Walden Pond in the world and distiller of celestial dews.

Who knows in how many unremembered nations' literatures this has been the Castalian Fountain or what nymphs presided over it in the Golden Age? It is a gem of the first water which Concord wears in her coronet.

Yet perchance the first who came to this well have left some trace of their footsteps. I have been surprised to detect encircling the pond, even where a thick wood has just been cut down on the shore, a narrow shelf-like path in the steep hill-side, alternately rising and falling, approaching and receding from the water's edge, as old probably as the race of man here, worn by the feet of aboriginal hunters, and still from time to time unwittingly trodden by the present occupants of the land. This is particularly distinct to one standing on the middle of the pond in winter, just after a light snow has fallen, appearing as a clear, undulating white line, unobscured by weeds and twigs, and very obvious a quarter of a mile off in many places where in summer it is hardly distinguishable close at hand. The snow reprints it, as it were, in clear white type alto-relievo. The ornamented grounds of villas which will one day be built here may still preserve some trace of this.

The pond rises and falls, but whether regularly or not, and within what period, nobody knows, though, as usual, many pretend to know. It is commonly higher in the winter and lower in the summer, though not corresponding to the general wet and dryness. I can remember when it was a foot or two lower, and also when it was at least five feet higher, than when I lived by it. There is a narrow sand-bar running into it, with very deep water on one side, on which I helped boil a kettle of chowder, some six rods from the main shore, about the year 1824, which it has not been possible to do for twenty-five years; and on the other hand, my friends used to listen with incredulity when I told them, that a few years later I was accustomed to fish from a boat in a secluded cove in the woods, fifteen rods from the only shore they knew, which place was long since converted into a meadow. But the pond has risen steadily for two years, and now, in the summer of '52, is just five feet higher than when I lived there, or as high as it was thirty years ago, and fishing goes on again in the meadow. This makes a difference of level, at the outside, of six or seven feet; and yet the water shed by the surrounding hills is insignificant in amount, and this overflow must be referred to causes which affect the deep springs. This same summer the pond has begun to fall again. It is remarkable that this fluctuation, whether periodical or not, appears thus to require many years for its accomplishment. I have observed one rise and part of two falls, and I expect that a dozen or fifteen years hence the water will again be as low as I have ever known it. Flints' Pond, a mile eastward, allowing for the disturbance occasioned by its inlets and outlets, and the smaller intermediate ponds also, sympathize with Walden, and recently attained their greatest height at the same time with the latter. The same is true, as far as my observation goes, of White Pond.

This rise and fall of Walden at long intervals serves this use at least; the water standing at this great height for a year or more, though it makes it difficult to walk round it, kills the shrubs and trees which have sprung up about its edge since the last rise, pitch-pines, birches, alders, aspens, and others, and, falling again, leaves an unobstructed shore; for, unlike many ponds and all waters which are subject to a daily tide, its shore is cleanest when the water is lowest. On the side of the pond next my house, a row of pitch pines fifteen feet high has been killed and tipped over as if by a lever, and thus a stop put to their encroachments; and their size indicates how many years have elapsed since the last rise to this height. By this fluctuation the pond asserts its title to a shore, and thus the *shore* is *shorn*, and the trees cannot hold it by right of possession. These are the lips of the lake on which no beard grows. It licks its chaps from time to time. When the water is at its height, the alders, willows, and maples send forth a mass of fibrous red roots several feet long from all sides of their stems in the water, and to the height of three or four feet from the ground, in the effort to maintain themselves; and I have known the high-blueberry bushes about the shore, which commonly produce no fruit, bear an abundant crop under these circumstances.

Some have been puzzled to tell how the shore became so regularly paved. My townsmen have all heard the tradition, the oldest people tell me that they heard it in their youth, that anciently the Indians were holding a pow-wow upon a hill here, which rose as high into the heavens as the pond now sinks deep into the earth, and they used much profanity, as the story goes, though this vice is one of which the Indians were never guilty, and while they were thus engaged the hill shook and suddenly sank, and only one old squaw, named Walden, escaped, and from her the pond was named. It has been conjectured that when the hill shook these stones rolled down its side and became the present shore. It is very certain, at any rate, that once there was no pond here, and now there is one; and this Indian fable does not in any respect conflict with the account of that ancient settler whom I have mentioned, who remembers so well when he first came here with his divining rod, saw a thin vapor rising from the sward, and the hazel pointed steadily downward, and he concluded to dig a well here. As for the stones, many still think that they are hardly to be accounted for by the action of the waves on these hills; but I observe that the surrounding hills are remarkably full of the same kind of stones, so that they have been obliged to pile them up in walls on both sides of the railroad cut nearest the pond; and, moreover, there are most stones where the shore is most abrupt; so that, unfortunately, it is no longer a mystery to me. I detect the paver. If the name was not derived from that of some English locality,—Saffron Walden, for instance,—one might suppose that it was called, originally, *Walled-in* Pond.

* * * *

The shore is irregular enough not to be monotonous. I have in my mind's eye the western indented with deep bays, the bolder northern, and the beautifully scolloped southern shore, where successive capes overlap each other and suggest unexplored coves between. The forest has never so good a setting, nor is so distinctly beautiful, as when seen from the middle of a small lake amid hills which rise from the water's edge; for the water in which it is reflected not only makes the best foreground in such a case, but, with its winding shore, the most natural and agreeable boundary to it. There is no rawness nor imperfection in its edge there, as where the axe has cleared a part, or a cultivated field abuts on it. The trees have ample room to expand on the water side, and each sends forth its most vigorous branch in that direction. There Nature has woven a natural selvage, and the eye rises by just gradations from the low shrubs of the shore to the highest trees. There are few traces of man's hand to be seen. The water laves the shore as it did a thousand years ago.

A lake is the landscape's most beautiful and expressive feature. It is earth's eye; looking into which the beholder measures the depth of his own nature. The fluviatile trees next the shore are the slender eyelashes which fringe it, and the wooded hills and cliffs around are its overhanging brows.

Standing on the smooth sandy beach at the east end of the pond, in a calm September afternoon, when a slight haze makes the opposite short line indistinct, I have seen whence came the expression, "the glassy surface of a lake." When you invert your head, it looks like a thread of finest gossamer stretched across the valley, and gleaming against the distant pine woods, separating one stratum of the atmosphere from another. You would think that you could walk dry under it to the opposite hills, and that the swallows which skim over might perch on it. Indeed, they sometimes dive below the line, as it were by mistake, and are undeceived. As you look over the pond westward you are obliged to employ both your hands to defend your eyes against the reflected as well as the true sun, for they are equally bright; and if, between the two, you survey its surface critically, it is literally as smooth as glass, except where the skater insects, at equal intervals scattered over its whole extent, by their motions in the sun produce the finest imaginable sparkle on it, or, perchance, a duck plumes itself, or, as I have said, a swallow skims so low as to touch it. It may be that in the distance a fish describes an arc of three or four feet in the air, and there is one bright flash where it emerges, and another where it strikes the water; sometimes the whole silvery arc is revealed; or here and there, perhaps, is a thistle-down floating on its surface, which the fishes dart at and so dimple it again. It is like molten glass cooled but not congealed, and the few motes in it are pure and beautiful like the imperfections in glass. You may often detect a yet smoother and darker water, separated from the rest as if by an invisible cobweb, boom of the water nymphs, resting on it. From a hill-top you can see a fish leap in almost any part; for not a pickerel or shiner picks an insect from this smooth surface but it manifestly disturbs the equilibrium of the whole lake. It is wonderful with what elaborateness this simple fact is advertised,—this piscine murder will out,—and from my distant perch I distinguish the circling undulations when they are half a dozen rods in diameter. You can even detect a water-bug (*Gyrinus*) ceaselessly progressing over the smooth surface a quarter of a mile off; for they furrow the water slightly, making a conspicuous ripple bounded by two diverging lines, but the skaters glide over it without rippling it perceptibly. When the surface is considerably agitated there are no skaters nor water-bugs on it, but apparently, in calm days, they leave their havens and adventurously glide forth from the shore by short impulses till they completely cover it. It is a soothing employment, on one of those fine days in the fall when all the warmth of the sun is fully appreciated, to sit on a stump on such a height as this, overlooking the pond, and study the dimpling circles which are incessantly inscribed on its otherwise invisible surface amid the reflected skies and trees. Over this great expanse there is no disturbance but it is thus at once gently smoothed away and assuaged, as. when a vase of water is jarred, the trembling circles seek the shore and all is smooth again. Not a fish can leap or an insect fall on the pond but it is thus reported in circling dimples, in lines of beauty, as it were the constant welling up of its fountain, the gentle pulsing of its life, the heaving of its breast. The thrills of joy and thrills of pain are undistinguishable. How peaceful the phenomena of the lake! Again the works of man shine as in the spring. Ay, every leaf and twig and stone and cobweb sparkles now at mid-afternoon as when covered with dew in a spring morning. Every motion of an oar or an insect produces a flash of light; and if an oar falls, how sweet the echo!

In such a day, in September or October, Walden is a perfect forest mirror, set round with stones as precious to my eye as if fewer or rarer. Nothing so fair, so pure, and at the same time so large, as a lake, perchance, lies on the surface of the earth. Sky water. It needs no fence. Nations come and go without defiling. It is a mirror which no stone can crack, whose quicksilver will never wear off, whose gilding Nature continually repairs; no storms, no dust, can dim its surface ever fresh; —a mirror in which all impurity presented to it sinks, swept and dusted by the sun's hazy brush,—this the light dustcloth,—which retains no breath that is breathed on it, but sends its own to float as clouds high above its surface, and be reflected in its bosom still.

A field of water betrays the spirit that is in the air. It is continually receiving new life and motion from above. It is intermediate in its nature between land and sky. On land only the grass and trees wave, but the water itself is rippled by the wind. I see where the breeze dashes across it by the streaks or flakes of light. It is remarkable that we can look down on its surface. We shall, perhaps, look down thus on the surface of air at length, and mark where a still subtler spirit sweeps over it.

The skaters and water-bugs finally disappear in the latter part of October, when the severe frosts have come; and then and in November, usually, in a calm day, there is absolutely nothing to ripple the surface. One November afternoon, in the calm at the end of a rain storm of several days' duration, when the sky was still completely overcast and the air was full of mist, I observed that the pond was remarkably smooth, so that it was difficult to distinguish its surface; though it no longer reflected the bright tints of October, but the sombre November colors of the surrounding hills. Though I passed over it as gently as possible, the slight undulations produced by my boat extended almost as far as I could see, and gave a ribbed appearance to the reflections. But, as I was looking over the surface, I saw here and there at a distance a faint glimmer, as if some skater insects which had escaped the frosts might be collected there, or, perchance, the surface, being so smooth, betrayed where a spring welled up from the bottom. Paddling gently to one of these places, I was surprised to find myself surrounded by myriads of small perch, about five inches long, of a rich bronze color in the green water, sporting there and constantly rising to the surface and dimpling it, sometimes leaving bubbles on it. In such transparent and seemingly bottomless water, reflecting the clouds, I seemed to be floating through the air as in a balloon, and their swimming impressed me as a kind of flight or hovering, as if they were a compact flock of birds passing just beneath my level on the right or left, their fins, like sails, set all around them. There were many such schools in the pond, apparently improving the short season before winter would draw an icy shutter over their broad skylight, sometimes giving to the surface an appearance as if a slight breeze struck it, or a few rain-drops fell there. When I approached carelessly and alarmed them, they made a sudden plash and rippling with their tails, as if one had struck the water with a brushy bough, and instantly took refuge in the depths. At length the wind rose, the mist increased, and the waves began to run, and the perch leaped much higher than before, half out of water, a hundred black points, three inches long, at once above the surface. Even as late as the fifth of December, one year, I saw some dimples on the surface, and thinking it was going to rain hard immediately, the air being full of mist, I made haste to take my place at the oars and row homeward; already the rain seemed rapidly increasing, though I felt none on my cheek, and I anticipated a thorough soaking. But suddenly the dimples ceased, for they were produced by the perch, which the noise of my oars had scared into the depths, and I saw their schools dimly disappearing; so I spent a dry afternoon after all.

Walden Revisited

George F. Whicher

Walden Pond, in spite of the twentieth century mania for calling every body of fresh water a lake, still keeps its old-fashioned name. Beside the heap of stones that marks the site of Thoreau's cabin the young pines are springing up. Standing there in the hush of an early morning in summer, one may look across the unruffled surface to wooded shores and recapture momentarily the sense of a landscape attuned to leisurely living and the delights of meditation.

But the drive and grind of modernity are not far distant. A glance at the ground where bottle caps and torn papers mingle with the dry pine needles or a whiff, not of a chance traveler's pipe, but of a passing motor-car's exhaust brings back the insistent realities of an era of speed and waste. Though Walden Pond is preserved as a State Reservation, it is no longer a refuge of the spirit. Free enterprise has exploited its bathing beach as a location for hot-dog stands and amusement booths. Official tests of the water at the beaches of popular resorts within the reach of Greater Boston show that Walden sometimes ranks among the most favored in urine-content.

Of more concern than the state of Walden's water or the beauty of its shores is the moral heritage of human freedom and loyalty to principle that descends from Thoreau. Nothing could be more opposite to the totalitarian doctrines of our times than the transcendentalist's belief in the dignity of man and the supremacy of individual conscience over a debased collective authority. The hope that these ideas may still be cherished and made valid is the mainspring of this essay.

Originally published as the Preface to *Walden Revisited*, by George F. Whicher (Chicago, 1945).

Walden in 1960

Edward Weeks

Twice in the course of the past three years, as I have reported in these columns, attempts have been made to desecrate Walden, one of the most beautiful great ponds in New England and a permanent memorial to Henry Thoreau. The first encroachment occurred when the Middlesex County commissioners, encouraged by the citizens of Concord, cut down three hundred yards of fine trees and bulldozed the topsoil into the water to form a raw, sloping hardpan convenient for young bathers. The second was when the authorities of Concord, in search of a new site for the town dump, finally dug the odoriferous hole at a distance less than half a mile from the reservation.

Concord is a town more conscientious than most; its citizens and Historical Society have been vigilant in preserving its Revolutionary and literary remains. Yet its attitude toward Walden seems that of an exasperated guardian, and part of the exasperation is doubtless caused by the picnickers (only a few of whom come from Concord), who follow that common American trait of throwing all rejected tin cans and hardware into the nearest available stream. (I remember seeing in one of the loveliest stretches in Upper Connecticut not only an iron bedspring under water but also the top half of an upholstered leather chair.) Walden has long suffered from the population pressure of suburbia, and I suppose that it was only human nature for Concord's citizens to give first thought to their own privacy and to leave the pond to the mercy of the tourists.

But the county commissioners should have know better, and I am

From *The Atlantic Monthly*, 206:92 (July 1960). Originally published in the column titled "The Peripatetic Reviewer."

pleased to say that in a recent ruling of the Massachusetts Supreme Judicial Court they have been ordered to halt their bulldozers and to replant trees felled at the site of Thoreau's cabin. The concrete ramps must be uprooted; a modern roadway must be restored to the semblance of a forest path; and one hopes that the hot dog stands will go too.

In short, the court found that the commissioners, in enlarging the recreation area, were not acting in accordance with the terms of the original gift made by the grandsons of Emerson when they presented Walden and the woods to the commonwealth. More than ten thousand dollars has been expended by the Thoreau Society in the legal fight for this victory, and it is heartening to know that the dollars, dimes, and pennies came from people the world over for whom Thoreau is still alive.

Cleaning Up the Merrimack

Leonard Wolf

The Merrimack River rises in the White Mountains of north-central New Hampshire and flows south, down the center of New Hampshire into Massachusetts; at Lowell, Massachusetts, it turns east and flows through Lawrence, Haverhill, and Newburyport to the Atlantic. It drains 5,000 square miles of New England and receives 13 major tributaries. A young river as the earth goes (probably less than 10,000 years old), it is old as the nation goes. It was settled early in the seventeenth century as part of the Massachusetts Bay Colony, and its swift flow was put to work running saw mills and grist mills. The Merrimack falls from an elevation of about 4,600 feet in a relatively short distance, and the force of its fall was ample to turn the heavy water wheels that ran mechanical looms before there was steam power. Soon after 1800 the Merrimack towns—Manchester and Nashua, New Hampshire; Lowell, Lawrence, and Haverhill, Massachusetts—became the centers of the U.S. textile industry, perhaps the first American company towns.

The Merrimack powered a great industry, and she paid dearly for it. The river is now thoroughly polluted. Once the mother of a great fishing industry, today she smothers her children. Fish require water with substantial quantities of dissolved oxygen in order to live and propagate. For the entire Merrimack the average oxygen reading is below the level required by most species for propagation; below Lowell there is frequently no oxygen in the water at all. Fish that manage to find oxygen to breathe may die of starvation, for the settling of waste solids covers the river bottom, destroying the small plants and animals

From *Bulletin of the Atomic Scientists*, 21:16-22 (April 1965).

that fish feed on. The clean, pebbly bottom that many fish (particularly the delicate species, such as salmon, that live in the sea but swim upstream to propagate in fresh water) require for survival is coated with ooze and sludge. The surviving fish of the Merrimack, predominantly the coarser types such as bullhead, shiners, and suckers, are tainted with oils, phenols, and dyes, and are considered unedible.

The shellfish industry at the mouth of the Merrimack is gone. Densities of bacteria and viruses in the water are so high that shellfish taken from these waters, if eaten raw, can cause serious illness, and the Massachusetts Department of Public Health closed the shellfish beds in 1926.

Any form of recreation—even boating or waterskiing—is hazardous on the Merrimack. The untreated municipal sewage which is poured into the Merrimack contains huge quantities of pathogenic organisms which can cause gastrointestinal diseases, such as typhoid fever, dysentery, and diarrhea; hepatitis; eye, ear, nose, and throat disorders; and skin infections. Even in the cleaner spots, for example at Tyngsborough Bridge near the Massachusetts-New Hampshire line, the sewage is so dense in the river that spray thrown up over a motor boat's windshield will expose a passenger to dangerous numbers of bacteria (about 42 in a single drop of water). The bacterial counts (measured in terms of one indicator organism, coliforms) exceeded 1,000,000 per 100 milliliters below several large cities (100 ml. is about half a cup). For comparison, 70 bacteria per 100 ml. is the maximum allowed for the production of safe shellfish, and 1,000 per 100 ml. the usual maximum for safe water recreation.

The swimmer or boater on the Merrimack may come into contact with highly toxic pesticides and other organic chemicals, such as phenols; toxic wastes from metal-plating and chemical industries have been observed to move from Leominster, Massachusetts, far inland, all the way to the river mouth; cyanide has been found, notably, very near the intake point for the City of Lowell water supply.

Having destroyed the fish-breeding and recreational potential of the river, pollution is now threatening the usefulness of the river as a source of municipal and industrial water. Detergents, particularly in the shallower tributaries of the Merrimack, cause sudsy water to run from faucets and can interfere with many industrial processes. Industrial wastes, particularly oils and organic chemicals, frequently leave objectionable tastes and odors, even color, in drinking water. Women from the Merrimack towns have complained that their laundry becomes yellowed. High ammonia concentrations in the water combine with chlorine in such a way as to reduce the effectiveness of disinfection processes, making it more likely that bacteria will survive in drinking water.

We are approaching the day, it would seem, when it is dangerous to live near such a river. It has long ceased to be pleasant. The stream is foul and smelly during the summer, when high temperatures accelerate the rot of organic wastes. Suspended solids settle onto the bottom of the river and decompose, sending gases bubbling up to the top as if the river were cooking. The gases often contain hydrogen sulfide, which smells like rotten eggs and can cause expensive damage by discoloring the paint on houses and boats. The river is completely opaque and usually black in color, though it is occasionally gray, milky white, or colored, depending on the particular variety of industrial wastes that are currently predominant. Oil slicks, grease balls, black oozing sludge, slimes, fecal debris, and condoms may float by on top of the stream.

The Merrimack did not turn into such an ugly and dangerous creature overnight. It was perhaps the earliest polluted river of our history. The earliest saw and grist mills discharged their byproducts—sawdust and grain chaff—into the river; the textile mills discharged wash-water, cotton fibers, wool greases, and the sanitary wastes from all their employees into the river. The needs of the textile industry and its workers brought other water-using industries to the Merrimack: machine building, tool and die manufacturing, wool scouring, silk preparation, dyeing, tanning, shoe manufacturing, paper production, and food processing. For all these industries and for the towns in its basin the river provided water power, process water, cooling water—and a free sewer.

A river makes a rather efficient sewer, since it can provide a degree of natural sewage treatment, in the form of aerobic digestion, for the wastes as it carries them downstream. The Merrimack, however, was taxed in a short time far beyond its abilities. In 1839, when Henry David Thoreau and his brother spent a week rowing down the Concord and up the Merrimack, the pollution of the lower Merrimack was already notorious. "When at length it has escaped from under the last of the factories," Thoreau wrote, "it has a level and unmolested passage to the sea, a mere *waste water*, as it were, bearing little with it but its fame . . ." (Emphasis in the original.) In the upper reaches above Lowell the river was still pleasant: cool, abounding with fish, a favorite spot for retreat. Thoreau listed the species of fish found here: freshwater sunfish; perch, trout and dace, already becoming rare; shiners; pickerel; horned pout; eels and lamprey eels; suckers; and, very occasionally, salmon and shad. He mourned the destruction of fish habitat and the idling of comercial fishermen:

> Perchance, after a few thousands of years, if the fishes will be patient, and pass their summers elsewhere, meanwhile, nature will have levelled the Billerica dam, and the Lowell

> factories, and the Grass-ground [Concord] River run clear again, to be explored by new, migratory shoals . . . One would like to know more of that race, now extinct, whose seines lie rotting in the garrets of their children, who openly professed the trade of fishermen, and even fed their townsmen creditably . . .

One hundred and twenty-five years later the Merrimack is much worse. In 1964 Thoreau would have had no desire to spend a week rowing on it. And the many attempts and promises to clean up the Merrimack have, so far, been unavailing. The history of Massachusetts' efforts in behalf of the Merrimack is illuminating—although New Hampshire's would do just as well.

The Massachusetts Department of Public Health, beginning in the nineteenth century, has conducted numerous studies of the river. In 1887 the department expressed the opinion that the advisability of using Merrimack water without treatment for drinking, as Lowell and Lawrence residents then did, might be reconsidered. After many more surveys the Department in 1913 recommended that cities treat their wastes—or else build a sewer from Lowell 40 miles to the ocean. [In] 1924 more recommendations for action were made; in 1928 another study followed, and in 1929 came another report with the same recommendations. State legislation from 1935 to 1938 created a Merrimack River Valley Sewerage District and Board, but no funds were ever appropriated for sewage treatment. In 1936 and 1938 the federal Works Progress Administration made studies and reports calling for corrective action. In 1945 the Merrimack River Valley Sewerage Board and the Department of Health published a joint report. But while the Merrimack's pollution has been bemoaned for nearly eighty years, in early 1964 not one town on the main stem of the Merrimack treated its wastes.

Massachusetts' failure to clean up the Merrimack has some good excuses. For one, the Massachusetts Department of Public Health had no enforcement powers whatever until 1945. Until that time, it could only make recommendations to the legislature, and that body, dominated by representatives of industries and reluctant taxpayers, was not ready to insist on large expenditures for pollution control. Even now the Massachusetts pollution control law provides no "carrot" in the form of tax incentives or grants for waste treatment works. A further difficulty is the current economic depression in the Merrimack Valley. Most of the Massachusetts communities are classified as areas of particularly high unemployment under the Area Redevelopment and Accelerated Public Works programs. Many of the textile industries, which no longer need water power for energy, have departed for the south and middle west; other old plants found it profitable to move into brand new facilities elsewhere rather than renovate outmoded equipment. Many remaining industries let it be known that sewer rates would be the straw to break their back, and force them out of business or into another state.

Industrial executives, rich or poor, are not terribly eager to build expensive waste treatment facilities, and they are not inexperienced in applying political pressure. The most well-intentioned officials may be brought to heel when industrialists are controllers of large chunks of the local economy or large shares of a party's budget. State health officials have upon occasion admitted that they have difficulty persuading their attorneys-general to initiate litigation against industrial polluters. Public opinion, mobilized by conservationist and civic action groups (the League of Women Voters is a great friend of pollution control officials), can be the key to persuading industries that pollution is unpopular and citizens that it is dangerous. On the Merrimack, where many residents feel themselves part of a dangerously depressed area, the citizens' campaigns were less effective. Perhaps there was a New England canniness, too, about public projects that would have to be built on credit and at great cost. In 1947, a bond issue for sewage plant construction was defeated everywhere except in Newburyport (the downstream residents and naturally the most anxious to clean up the river) by tremendous margins. In Haverhill the vote was 15,385 to 3,627, in Lowell 37,166 to 5,087.

If there was ever a need for federal action it was here on the Merrimack, and since 1956 there has been legislation giving the federal government an active responsibility for cleaning up the nation's waterways. Under this federal Water Pollution Control Act, the Secretary of Health, Education and Welfare is given enforcement authority over interstate pollution, and $100,000,000 is authorized annually for 30 per cent grants to towns for constructing waste treatment facilities. Although administration of the act rests with the Public Health Service, under Surgeon-General Luther Terry, amendments to the act in 1961 put policymaking responsibility for enforcement in the hands of the Secretary alone. The Merrimack clearly falls within his enforcement jurisdiction. The Secretary is directed to initiate enforcement proceedings when he has reason to believe "on the basis of reports, surveys, or studies" that there is interstate pollution damaging to health or welfare. The Merrimack is interstate, it is polluted, and the department's staff had been informed of its condition by repeated reports, surveys, and studies. The act also states that the federal government is to "recognize, preserve, and protect the primary responsibilities and rights of the states in preventing and controlling water pollution." On the Merrimack the inability of the Commonwealth of Massachusetts to prevent and control pollution had been demonstrated over eighty years. The then junior senator from Massachusetts, John Kennedy, and several Massachusetts congressmen had requested federal action soon after the act was passed. Yet none was taken until eight years later.

This neglect of a crucial pollution problem is no exception in the history of the federal water pollution control program. The Kennedy-Johnson administration, pledged to fight pollution as top priority in a strong conservation program, initiated only 17 enforcement actions; this is not what one would call a vigorous regulatory program, considering that the slow-moving Eisenhower administration produced 13 enforcement cases and that the department itself has promised to act on some 90 cases of serious interstate pollution. (The total of 30 federal water pollution cases does not include such key polluted rivers as the Ohio, the Delaware, the Susquehanna, the Hudson, or the Arkansas.) The less-than-glowing record of the department in this program is due, in part, to the fact that the Public Health Service does have a great deal of influence over policy in the water pollution program. Since the nineteenth century, the Public Health Service has been engaged with state and local officials in an advisory capacity, and has won great victories in sanitation, controlling epidemics, free vaccinations, and the like by persuasion and cooperation. Organized like the Army, Public Health Service officers have military ranks, a strong esprit de corps, and a great desire not to make enemies. The service has not been able to adapt easily to the role of a regulator—a role which practically requires making enemies—both because of its outlook and experience and because so many of its other activities still depend on cooperation and good feeling. Pollution control, particularly in New England and in the south, tends to become a states' rights issue, and state authorities see federal intervention as a punitive measure. (Other, more enlightened, officials have successfully used the threat of federal intervention to their great advantage as a buck-passing device.) Surgeon-General Terry has not been anxious to use federal enforcement in cases where state health officials who have come to trust him are, in his opinion, doing their best.

Apparently this was the case on the Merrimack, where it was revealed that a Public Health Service official promised the Massachusetts authorities that federal enforcement would not be initiated.

* * * *

When federal enforcement finally came to the Merrimack in 1964, it was not at the initiative of the Public Health Service, but at the request of the newly elected governor, Endicott Peabody. (Under the federal act, the Department of Health, Education and Welfare is obliged to act if requested to do so by a governor.) On February 12, 1963, in a letter to Secretary Celebrezze, he formally requested an enforcement conference on the Merrimack, its interstate tributary the Nashua, and their smaller tributaries in Massachusetts and New Hampshire. The governor, new to the job, may have underestimated the amount of opposition he would evoke. There were some rumors and newspaper accounts saying that he was anxious to withdraw his request. The conference was postponed for a year, so that an engineering study authorized by the Massachusetts legislature could be completed before the conference. The Public Health Service received the report of that study in December 1963. When the Public Health Service applied to the Massachusetts Department of Public Health for more data, they were refused, the Massachusetts authorities refusing to name specific industrial polluters. Only after Public Health Service engineers went to Massachusetts, making it clear that they would collect the needed information themselves if necessary, did the Massachusetts Department of Public Health authorities capitulate.

The state of New Hampshire challenged the legality of the entire conference. With the health authorities, the state's assistant attorney-general came to the conference to file a reservation of rights, agreeing to participate without acknowledging the jurisdiction of the Secretary of Health, Education and Welfare.

At the conference the opponents of federal "intervention" outnumbered the supporters both in numbers and vehemence. State Senator William X. Wall said: "These evangelists for purification of the Merrimack River should realize the prohibitive cost entailed . . . I am not against cleaning up the Merrimack River . . . I have fought for this thing fifteen years ago. But I am against saddling the taxpayers with a Frankenstein debt running into millions of dollars, and I sincerely hope that Uncle Sam comes up and gives us the money to do it." Cornelius T. Finnegan, Lowell city solicitor: "I have been delegated by the city manager of the city of Lowell, by a vote of the City Council, to oppose any action and to protest any action which would require the expenditure of money by the city of Lowell." Donald M. Crocker, chairman of the Nashua River Committee of the National Council for Stream Improvement: "Thus, while the Nashua River is an interstate stream it cannot be said that the use of the river by Massachusetts paper mills endangers the health of persons living in either Massachusetts or New Hampshire." (A true statement, since paper wastes are not usually high in pathogenic bacteria. It neglects to mention the countless other damages created by paper wastes—such as objectionable odors and color, slimes and fungi, destruction of dissolved oxygen supplies.) Lowell City Manager P. Harold Ready: "If there is a great demand for clean water it hasn't yet reached the Merrimack Valley. I can't see the urgent necessity for speed in this thing." The *Lawrence Eagle-Tribune*, editorializing: "Purification is in the class of a Rolls Royce or an ermine coat—something nice to have but something we don't need . . ." Andrew Gillis, ex-mayor of Newburyport and understandably angry that Newburyport had spent about $1,110,000 to build a sewage treatment plant that would do no good since no upstream town had treated its wastes, said: "I have been against the Merrimack River pollution program since it started . . . And you would think down in Washington that Washington was giving you that dough. You would think Washington was. I wonder where that dough comes from. It comes from suckers like you and I here, and Mr. French [Lowell industrialist], the man that's out trying to make a dollar. It ain't coming from these goldbrickers here that are getting paid by the federal government and the state of Massachusetts."

Industry representatives tended to express their views rather obliquely. For example, the general manager of a combing plan in Massachusetts explained:

> The company employs approximately 700 people, with an annual payroll in excess of $2,500,000. This is new money from outside which is pumped directly into the Greater Lowell business stream each week. . . .
>
> We pay local real estate taxes of about $33,000. Wherever we can purchase or subcontract locally, we do so. . . .
>
> Pleasure activities such as boating, swimming, and fishing along the rivers and natural waterways are important. Industry is also certainly important. We should give careful, deliberate thought to a proper balance between these things. We should keep well in mind the fact that if you want to maintain a prosperous economy, it is industry in this area which must do it.

* * * *

The representative of the Department of Health, Education and Welfare who chairs water pollution conferences is Murray Stein, chief enforcement officer of the water pollution control program. Stein usually tries to get unanimous conclusions and recommendations out of the conferees, but did not prevail at this conference, to no one's surprise. Neither state would admit the occurrence of interstate pollution, both advancing the somewhat imaginative proposition that all wastes discharged in New Hampshire remained in that state despite the flow of the river. Both New Hampshire and Massachusetts engage in a procedure of "classifying" rivers, that is deciding how clean they should be for the uses customarily made of them, before prescribing requirements for treatment. New Hampshire said it could not classify the Merrimack until 1971, after which it would embark on a 12-year program to have adequate treatment facilities by 1983. Massachusetts said that unknown factors made it impossible to offer any schedule for action at that time; subsequent to the conference Massachusetts authorities promised to classify their portion of the Merrimack in 1965 and to require preliminary plans for sewage treatment plants in most cities by September 1, 1964. The federal conferee recommended that construction be begun on waste treatment facilities for all places needing them by May 1, 1967.

On June 11, 1964, Secretary Celebrezze wrote to the conferees, sending copies of the official summary of the conference, and recommending that the Massachusetts schedule of action be carried out. This means that Merrimack towns should now all be working on development of final engineering plans and financing arrangements for sewage treatment facilities. Under the law, the Secretary's recommendation has a definite legal weight. If, at least six months later, adequate remedial action has not been forthcoming, the Secretary is instructed to call a formal hearing, the second stage in enforcement action. If the hearing board also finds that adequate remedial action is not being taken, it may again issue recommendations; if there is still no compliance after at least six months, the Secretary may request the Attorney-General to file suit against the polluters. In the history of the federal Water Pollution Control Act, only one such suit has been filed and only four hearings called. The enforcement officers feel this is a fine record, since they prefer to see the abatement of water pollution occur as a result of cooperative federal-state action.

The cleanup of the Merrimack may be difficult to arrange in a spirit of cooperation, however. It has been polluted for so long that the possibilities of its being more than a sewer are not always obvious. "Merrimack Valley people have been living with the river for generations. . . . They can continue to live with it indefinitely without fear their health will be harmed," said the *Lawrence Eagle-Tribune*. Opposition grew after the conference recommendations were issued. Lowell City Manager P. Harold Ready called for legislation to diminish the powers of the sanitary engineering division of the state Department of Public Health, and he was promised support by industrialists and local state representatives. The fear of losing industries made most towns offer to permit industrial wastes to be discharged into city sewers, with treatment to be supplied by municipal plants. Industrial wastes can be an enormous burden to a town sewage system. In one small town alone (Fitchburg, Massachusetts), three paper plants discharged solids equal to the raw sewage from a population of 408,000. The wool scouring operations in Chelmsford and Lawrence contribute over three tons of grease daily. It is estimated that 15 per cent of the unnatural color of the river at Lowell has come from the Franconia Paper mill at Lincoln, New Hampshire, 131 miles upstream. Furthermore, certain industrial wastes —greases, metals, and organic chemicals, for example—require types of treatment different from that usually applied to domestic wastes. The task of financing falls on the cities at a time when Accelerated Public Works and Area Redevelopment grants are no longer available, the 88th Congress having failed to appropriate funds for their continuation. The Water Pollution Control Act offers grants of up to 30 per cent, which the New Englanders consider too little, and the Commonwealth of Massachusetts to date offers no aid for this purpose.

The compliance schedule that Secretary Celebrezze set his name to called for preliminary engineering plans for treatment facilities for all the towns to be completed at various times between August 1964 and September 1965. Full schedules, including dates for the completion of construction are due three months after the submission of preliminary plans. It must be expected that construction will require a long time, especially since interceptors and sewers will need to be laid to connect industries and new communities. Sometime after 1970, however, if action is taken now, New England residents might expect to see results. And the results could be spectacular if good equipment is installed.

On the assumption that New Englanders might respond more to a dollars-and-cents argument than to an aesthetic one, some conservationists have tried to estimate the economic increments to be expected from clean water in the Merrimack. First of all, a profitable tourist trade could develop. Public Health Service experts believe that tasty freshwater fish, even trout and shad, would return naturally if the water were clean. The continued growth of all the eastern seaboard cities will provide customers for a whole range of water recreation facilities. Land along the river will increase in value, and an influx of sportsmen and vacationers would increase profits of all consumer service enterprises. Secondly, the river may well be needed as a source of cheap drinking water, which can be used over and over again. Daily per capita consumption of water in Nashua, New Hampshire, for example, was 155 gallons in 1963. It had doubled since 1910, and will double again in only 20 more years. Groundwater supplies will not be adequate for increased demand. Lawrence and Methuen, Massachusetts, already are obliged to take drinking water from the Merrimack; Lowell and Chelmsford are making plans to do so; and doubtless other towns will have to do so in the near future. If Merrimack water is to be used safely, it will have to be less contaminated than it is today. Third, a good steady water supply will be attractive to many industries. If the federal government is stern in enforcing pollution control nationally, it will be of little avail for industries to leave one neighborhood to avoid having to treat their wastes.

Life on a Dying Lake

Peter Schrag

North Central flight 940. We take off to the west from the Detroit Metro Airport, turn left in a large loop around the city, and emerge over the crotch of the Detroit River where the patterned effluent spills into the waiting blue water of Lake Erie. *Two weeks before the final count-down for Apollo II, the supreme $24-billion apotheosis of American technology; the greatest thing, Richard Nixon will say, since the creation.* The brown waters hug the western shore, and beyond, through the thin haze from the Michigan stacks, the ore carriers cut white wakes to feed the factories. We cross the rectilinear fields of Pelee Island and pass Pelee Point on the Ontario shore to the north. Less than 150 miles to the south is Wapakoneta, Ohio, where the first man to set foot on the moon was born.

> Man is destroying Lake Erie. [Says the report from the Federal Water Pollution Control Administration.] Although the accelerating destruction process has been inadvertent, it is as positive as if he had put all his energies into devising and implementing the means. After two generations the process has gained a momentum which now requires a monumental effort to retard. The effort must not only be basin-wide and highly coordinated; it must be immediate. Every moment lost in allowing the destruction to continue will require a longer, more difficult, and more expensive corrective action.

From *Saturday Review*, 5:18-21, 55-56 (September 20, 1969).

Prosaic language from Washington: eutrophication, secondary treatment, nutrient removal, algal blooms, biological oxygen demand. There is little romance in a sewage plant, and none in the technicalities of oxygen depletion, thermal stratification, or discharge of organic wastes. But the destruction of a lake is not merely a technical or a political problem. To think of it is to think of all America, of our love-hate relationship with our technology, about our ambivalence about who we are and what we are, about the Hudson and the Missouri, about the Santa Barbara Channel and nuclear bombs, about defoliation in Vietnam and DDT-poisoned fish in Michigan—about all the things we value, often in contradiction—in our past and our future.

As the plane descends over the murky waters along the Cleveland shore, the brown edging the blue, I have to imagine kids I know, kids like Corky Divoky years ago, scanning the skies for hawks and heron, or walking the ice in the winter, and now confronting the public servants of the draft board and the crusade for freedom in Asia. I hear the reminiscences of old men, telling fish stories about walleye and whitefish and blue pike, species that have all but disappeared. How much, you ask, setting down at Hopkins Airport, is this worth; how much is romance and sentimentality about a fading past, how much the price of progress, how much the comfortable guilt of safe men who can attack pollution as an undisputable evil (or war, or technology itself) while languishing in their benefits?

Some thirteen million people live in the basin of this lake, 90 per cent of them on the American side, the rest in Ontario, the polluters and the polluted, perpetrators and victims, all of them dependent on a body of water that, according to the best evidence, is not yet dead but in danger. They drink its water, swim on its beaches, eat its fish, and sail from its harbors. At the same time, they, their cities, and their factories each day dump, leak, pipe, or drop into the lake several hundred million pounds of sewage, chemicals, oil, and detergents fouling beaches, killing wildlife, and imperiling the water itself. Sometimes you can smell and taste it as it comes out of the tap, sometimes you can see it on the beaches and often in the rivers—the Maumee, the Auglaize, the Ottawa—but most significantly, you fear, not what already exists, but what might—and could happen—if the process continues.

In parts of the western end of the lake, the blue-green algae, which thrive on the excess of nutrients from sewage, turn blue water to a murky green and accumulate in heavy mounds on the shore; in Sandusky, once one of the largest freshwater ports in the world, a large fishing industry has been reduced to a couple of operators who truck their low-grade catch to Georgia (where, apparently, people are still hungry enough to buy it); and in Cleveland, the industrial stream called the

Cuyahoga River is pronounced a fire hazard, a declaration that sounded hyperbolic until, last June, the oil on the river began to burn, damaging two bridges.

Cleveland's two fireboats travel the river periodically, hosing oil off docks and pilings so that the inflammable ooze will slowly make its way downstream and into the lake. At the same time, a broken city main is dumping twenty-five million gallons of raw sewage into the river each day. Periodically, the main is repaired, and now the city, with a $100-million bond issue voted last fall, is preparing to improve collection and treatment for its entire system and for the neighboring communities that it serves. (Cleveland, incidentally, may also be one of the few municipalities in the world that chlorinates its lakefront beaches so they will be safe for swimming.) But when it comes to pollution of the lake, Cleveland is more a victim than a culprit. Cleveland fouls its own nest with its dirty rivers and its inadequate sewage system, while Detroit, which dumps the waste of a huge industrial population into the Detroit River, stocks Lake Erie. Approximately 65 per cent of the oxygen-depleting wastes in the lake come from Detroit; 9 per cent from Cleveland. "When it comes to polluting the main part of the lake," said a researcher at the U.S. Bureau of Commercial Fisheries, "Cleveland's hardly on the map."

And yet, in a way, everything is on the map. Everything contributes to, and suffers from, the condition of the lake: people in five states and a Canadian province—hundreds of towns and cities from Toledo to Buffalo: Akron, Erie, Cleveland, Lorain, Conneaut, Ashtabula. The federal government has identified 360 sources of industrial waste—power plants, steel mills, chemical companies, food processors, rubber companies. During every heavy rain, flooding sewers and silt spill into the lake, and even in normal periods silt and fertilizers and pesticides drain into its tributaries. But the greatest polluters may be the city sewage systems themselves. The federal government has estimated that with existing treatment facilities, the cities along the lake discharge effluent equal, in its composition and effects on the lake, to the raw sewage from a population of 4,700,000 people. Some cities are providing secondary treatment, some primary, some none at all. Lake Erie has been called a huge cesspool, an appellation that has at least marginal accuracy. What is absolutely accurate is the statement that in the past fifty years pollution has substantially altered the ecology of the lake, and that it has made the lake far older than its years.

The word, among the scientists, is eutrophication—the process of aging. All lakes grow old as they collect runoff and materials from the surrounding shores. Over thousands of years they eventually accumulate enough silt from erosion and organic materials to turn them into marshes and, finally, into dry land. In Lake Erie man has accelerated that process with his wastes and sewage. An excess of nutrients, primarily phosphates and nitrates, has produced great growths of algae in the water and impaired the oxygen supply, especially in the deeper water during the summer—and especially at the western end, which is hit hardest by the excrement from Detroit. (Biological degrading of the nutrients requires oxygen; when the nutrients are too heavy the oxygen becomes depleted.)

Mayflies, which once grew in huge numbers in the western, the shallowest, end of the lake, and which provided a food supply for fish—cisco, blue pike, walleye, and other species—have declined; the water has been taken over by sheepshead, carp, and other types that are tolerant of low-oxygen conditions and whose eggs can survive the accumulation of sediments at the bottom. Some species that have surmounted changes in food supply and depleted oxygen now take longer to reach maturity. (The total volume of fish caught in the lake each year is as large as ever, but the catch is worth only half of what it was ten years ago; most of the fishing is now done by Canadians.)

On occasion there have been duck kills, flights of birds which have landed on oily water and never flew again, either (it is assumed) because the oil destroyed the birds' natural protection against the water, or because they were poisoned. (There has been some serious talk in recent years about oil drilling in the lake. So far the derricks of the Canadian Pacific Oil and Gas Company, which have been erected on the Ontario side of the lake, are producing only natural gas.) There is also a possibility that the algae, under certain conditions, can manufacture their own poisons, endangering wild life—and possibly human life.

Because the lake is relatively shallow, there is hope that once the rate of pollution is retarded (hopefully, but not certainly, through improved sewage treatment), the lake, with proper oxygen circulation, can recover, spilling its wastes into Lake Ontario and, ultimately, into the Atlantic. The certainty of that recovery and the effectiveness of the measures now being planned (which, among other things, include the removal of phosphates before effluent is discharged into the lake) are still matters of debate—and of time. What is not a matter of debate is that in the past fifty years, Lake Erie has aged 15,000 years.

Barry Commoner, a Washington University (St. Louis) biologist, who has long been concerned with the abuses of technology, writes:

> The lake is threatened with death. . . . Since the area was first settled, Lake Erie has been increasingly burdened with organic wastes and with inorganic nutrients that the lake's algae convert to organic materials. These organic materials would long ago have asphyxiated most of the lake's living

> things had it not been for the peculiar power of Iron III [an iron compound called ferric iron] to form insoluble complexes with the materials of the bottom mud. The protective skin of Iron III has held the enormous accumulation of potential oxygen-demanding material in the muddy bottom of the lake. But this protective skin can remain intact only so long as there is sufficient oxygen present in the water over the mud. For many years this was so, and the layer of Iron III held the accumulating mud materials out of the lake water. But a serious oxygen depletion now occurs in the summer months. As a result, the protective layer of Iron III has begun to break down—exposing the lake to heavy impact of the accumulated algal nutrient long stored in the mud. If the process continues, we may face a sudden biological cataclysm that will exhaust, for a time, most of the oxygen in the greater part of the lake water. Such a catastrophe would make the lake's present difficulties seem slight by comparison.

The fear is that under existing conditions, Erie could, without warning, turn into a huge swamp. Among the officials of the Federal Water Pollution Control Administration (FWPCA), which is charged with enforcing pollution control measures, Commoner is regarded as a prophet of gloom, a false Cassandra who is trying to frighten people. Nonetheless, FWPCA paraphrased Commoner's statement in its own report on the lake. "Some of that," said one FWPCA official, "was a little exaggerated. We know that the Iron III tends to break down and to release nutrients from the bottom, but there hasn't been any cataclysm, and there's not going to be one."

And yet, perhaps, that's not the issue—shouldn't be the issue. The trouble with conservation is that it has always been a matter of calamities and cataclysms. In the confusion of state, local, and federal antipollution responsibilities, there is always a large measure of sympathy for the company or the city that has to spend money for better treatment facilities, for the corporate taxpayer who might move somewhere else, for the time it takes, for the problems involved. The questions are thus always questions of resources, of priorities, of urgency and time. How much are fish worth? What's the value of a duck? What is the relationship between defoliants in Vietnam (or the price of automobiles in Detroit) and an acid discharge on the Cuyahoga or the Maumee?

If a major corporation can increase its earnings $60-million a year by raising prices, then is the expenditure of $18-million for new waste-treatment facilities at its plant in Cleveland something to brag about?

How much passion and effort are required by the hypothetical impairment of a municipal water supply? ("The time's going to come," said an angry conservationist in Cleveland, "when they won't be able to put any more chlorine into the drinking water. What the hell are they going to do then?") And to what extent is water or air pollution a problem that only the comfortable can afford? The kids from Cleveland's Hough ghetto (and many others) rarely worry about swimming from polluted beaches; for them and their parents, urban pollution has other, more virulent forms. Their problem, among other things, is rats, not fish.

In the meantime, life on the lake goes on. In most places the problem is invisible; in others it becomes part of the background, an element of lore, like a volcanic mountain on a South Sea island. "People come here expecting to see a swamp," said a Cleveland newspaper reporter. "But there isn't any." On the west side of Catawba Island, near Sandusky, the cabin cruisers and the yawls luxuriate in their elegant marinas, and on the northern tip the Chevies back their trailer-borne runabouts onto concrete ramps for a few hours of fishing or a cruise to Put-in-Bay.

* * * *

On the Cuyahoga River, twice each day, the excursion boat *Goodtime II* takes tourists on a run of the industrial sites; a tape-recorded spiel piped over the *Goodtime's* loud speakers describes the adjacent activities of the Great Lakes Towing Co., U.S. Steel, Republic Steel, Sherwin-Williams Paint, National Sugar Refining, and Standard Oil. (No word about waste discharges, about the phenols and oils and acids that ooze into the river. "If you fall in," they say along the river, "you won't drown, you'll decay.") On summer evenings at the foot of the river, on the flats off Front Street, the customers of Fagan's Beacon House sit at their tables just above the ooze listening to Dixieland, drinking beer, and watching the ships go by. It is, they say in Cleveland, one of the places "where everybody goes."

* * * *

The lake is life: Euclid Beach and Cedar Point, Catawba and Sandusky Bay, amusement parks, boats for hire, elegant summer houses tree-shielded from the curious roadway, beach-club privacy, and breakwaters spiny with the antennae of fishing poles. The beer cans collect between the rocks of the jettie, and behind them ragged rubber tires, twigs, and oil cartons undulate against the stones. A CLEAN BEACH IS A

FUN BEACH reads a sign posted two hundred yards from Cleveland's own outfall of sewage. "Why can't we have swimming in Lake Erie?" the mayor asks his commissioner of public utilities, and so the water around the beaches is fenced in with heavy sheets of Dacron anchored to the bottom, the chlorine is piped in, and the black workman rake the accumulated algae into little piles. On July 4 the beach is dedicated to safe swimming. "I have this vision," says a smart mouth. "The mayor arrives in a helicopter and climbs down a rope ladder onto the algae, while the Cleveland Orchestra plays 'Shifting Sands.'"

Who controls this environment? Whose rights are invested in it? The mayor—in this case Carl B. Stokes of Cleveland—is accused of being more interested in creating an image through waterfront chlorination projects than in attacking the fundamental problems. After the Cuyahoga fire, he begins legal action to make the state enforce its own anti-pollution standards against the industries on the river (and perhaps to put a little pressure on the corporate managers); the state, in turn, accuses the mayor and the city of lagging behind in their own anti-pollution efforts: "It is obvious," said an official of the Ohio Health Department, "that Cleveland itself is a major contributor to the problems of the Cuyahoga River and that a major cleanup cannot be accomplished until the city corrects its own faults."

All the states claim that their industries are in compliance with state health and pollution regulations, which simply means that each year they issue a few admonitions, ask the corporations what efforts they propose to make, and let them continue to operate. "All they're doing," someone said in Cleveland, "is licensing the polluters. It seems that it's impossible for anyone to be in violation." (To which the federal people reply that industry is making more progress with waste-treatment than the cities.)

The buck is supposed to stop at the federally conducted enforcement conferences. Under the club of federal authority, all the states in the lake region—Indiana, Ohio, Michigan, New York, and Pennsylvania—have committed themselves to upgrading their municipal waste-treatment facilities to the point where 80 per cent of all phosphates are removed from sewage effluent before it is discharged into the lake. (Some cities are already behind schedule and have received extensions.) Detroit and Cleveland, among others, are building new plants and collection facilities, using local, state, and a little federal money. At the same time, their officers are angry at the paltry federal contribution. While Detroit is scheduled to spend $159-million and Cleveland has voted its $100-million bond issue for better treatment and collection, the federal government spends barely more than $200-million a year on pollution research and development for the entire nation. "The federal people are the biggest hypocrites in the bunch," said a Detroit official. "They go around the country making speeches. Maybe if they made as much noise about getting us more money as they do about dirt, we'd be able to move a little faster." The reply from Washington: "The people who are polluting are responsible."

So far, the lake has been unaffected, and there is doubt that even after the scheduled projects are completed they will be sufficient. In New York State, a Health Department advisory committee of scientists was asked the question: Will phosphate removal retard eutrophication in Lake Erie? The answer, in simple words: Don't expect too much. Nonetheless, FWPCA has committed itself to the process as a necessary first step: phosphates, which come largely from detergents say the FWPCA technicians, are essential to the growth of algae and plankton, so (because it is relatively easy) phosphates will be removed. No one has yet figured out what to do with the phosphates once they are precipitated out. They will be trucked to . . . where? (Nor is it certain that nitrates, which flood into the lake from agricultural fertilizers, and other sources, are not major factors in eutrophication.)

"The scientists can raise all sorts of questions," says Murray Stein, who is charged with FWPCA's enforcement work. "It doesn't mean they aren't good questions, but every day's delay in studying means time lost forever. You can only hope." (In Ontario, the Water Resources Commission, which has its own problems, speaks about a "breakthrough" in sewage treatment—a chemical-physical process that is said to remove nutrients more effectively and cheaply than existing methods. But the process, so far, has only been tried under experimental, not operational, conditions.)

The FWPCA has estimated that it will cost $1.1-billion in pollution control projects to arrest the process of eutrophication in the next twenty years. Some critics believe the figure is far too low, and a few believe that the job is already impossible, that the lake may already be too far gone.

Who controls this environment? Whose rights are affected, whose life? The issue of pollution creates its own bureaucracies, its own inertia, its own zones of indistinct responsibility. Even though there are federal laws dating back to 1899 that prohibit the dumping of oil and refuse into navigable waters, and even though the various states have established their own regulations, there has never been—in the memory of federal officials—one suit or one criminal proceeding against a polluter. "Pollution law is a little like antitrust law," a federal official tells you. "It's hard to establish a connection between discharges and damage."

Whose rights are affected, whose environment is it? "Every year there's more talk," says David Blaushild, a Cleveland automobile dealer who heads a group called Citizens for Clean Air and Water. "The governor and the mayor come to the conferences, and make speeches, and go home, and the pollution goes on. You don't have to study anymore. You can smell and see it. It's time to file lawsuits. Why should people take this crap?"

Does the individual have a constitutional right to clean air and water? "It's going to take a disaster to wake people up," Blaushild says. "If this generation doesn't do it, the next generation won't know any better. They'll think that swimming in filth is the normal thing to do. They'll think the moon is supposed to be yellow. They'll think they're breathing clean air and drinking clean water, because they won't know any better." Blaushild writes the Cleveland science museums to ask why they don't take a stronger position against industrial polluters. (Among their trustees are directors of several local corporations.) One of them answers that Blaushild, by selling Detroit's products, has his own share of responsibility.

Who controls this environment? Pollution, pesticides, fallout. The world's experience with nuclear tests has begun to create a wholly new concept of civility and community. In a strange way pollution became a problem by analogy: we learned, for example, that the same ecological processes that concentrate strontium-90 in bones concentrate DDT in fish, that contamination in one place jeopardizes life in others. A bomb test in New Mexico kills infants in Mississippi and Alabama; pesticides on Michigan farms poison fish in distant lakes; sewage from Detroit fouls beaches in Ohio. One can respond cheaply by lamenting the fix that science and technology have gotten us into, but a bumper sticker proclaiming SAVE LAKE ERIE pasted barely a foot above a smoky automobile exhaust is more an illustration of the problem than a solution.

The burden of moral compromise symbolized by Hiroshima and Nagasaki will not be lifted by building a new sewer system in Detroit, however necessary that system may be. Technological amelioration of one facet of environmental destruction can be no more than a surrogate for continued acceptance of its larger and more catastrophic forms. Can one take seriously an organization whose interest in conserving fish is unmatched by a position on the ABM?

The questions are backwards: how much civility can we afford after we have paid for Vietnam? For the car? For our missiles? Can we sustain a decent welfare program despite the war? Can we clean the river without jeopardizing the profits of industry? Because we are trying to satisfy a new, though still unclear, sense of community with old priorities, evasion is inevitable. Which is to say that a professed commitment to protect an environment that ends with a squabble over sewer taxes is no commitment at all. The issue of pollution can produce a paranoid fanaticism just like every other; no one has died from swimming on a contaminated beach on Lake Erie or from drinking its water. Yet somehow, if we cannot distinguish between fanaticism on behalf of a distant generation and that which defends immediate returns and private ends we have simply lost our claim to live.

As Barry Commoner wrote in *Science and Survival:*

> The environment is a complex, subtly balanced system, and it is this integrated whole which receives the impact of all the separate insults inflicted by pollutants. Never before in the history of this planet has its thin life-supporting surface been subjected to such diverse, novel, and potent agents. I believe that the cumulative effects of these pollutants, their interactions and amplification, can be fatal to the complex fabric of the biosphere. And because man is, after all, dependent on part of this system, I believe that continued pollution of the earth, if unchecked, will eventually destroy the fitness of this planet as a place for human life.

If the greatest thing since the creation is worth twenty-four billion clams, how much is the creation worth?

Everybody Should Break an Ankle

Gene Marine

In a lawsuit in San Francisco a few years ago, an attorney was trying to recover damages for a workman who had broken his ankle on the job, and on the stand a doctor testifying for the insurance company was vigorously insisting that no permanent damage had been done. The ankle, the doctor insisted, was as good as new.

"In fact," the doctor argued, "the cartilage that has grown up at the point of fracture is actually stronger than the original bone."

The attorney turned to the jury, raised an eloquent eyebrow, and turned back to fix the doctor with a withering stare. "Tell me, doctor," he asked, "do you recommend this type of fracture to all your patients?"

All across the United States, the Engineers (and particularly the Engineers who work for private industry) are recommending fractures as improvements on the national ankles. In fact, if Engineers across the country are engaged in rapine, it is in private industry particularly that they find their public relations colleagues eager to assure the public that rape is really a therapeutic treatment designed to soothe anxiety, improve circulation and whiten the teeth. They are experts at finding ingenious ways to convince you that fractures are good for you.

Willard F. Cheley, assistant to the board chairman of the Georgia-Pacific Corporation, told *The New York Times* that "the corporation is so interested in conservation that it plants five trees for every one it chops down in its lumber business—sometimes it plants ten and fifteen trees. We are very conservation minded." That the trees cut down are

From *America the Raped*, by Gene Marine (New York, 1969), pp. 82-101.

sometimes in irreplaceable virgin Douglas fir stands, Mr. Cheley omits to say.

The Arcata Redwood Company—which did its best to cut down the redwoods along Redwood Creek before anybody could protect them in a park—posts signs along the highway when the ruinous consequences of its work cannot be hidden. They read: OVERMATURE TIMBER HARVESTED HERE. Overmature timber, of course, is any tree big enough to cut down and make a profit on.

Among the best of the fracture-sellers are the giant utility companies. The largest of them—California's Pacific Gas and Electric Company—not only spends huge amounts every year to keep its "image" clean (through such activities as a sizable annual contribution to the San Francisco Museum of Art, for instance), but "encourages" its executives in every northern California town to become unselfish civic leaders.

And then there's Con Ed.

Indian Point, on the Hudson River, is the site of a nuclear power station opened by Con Ed, with great fanfare, in 1963. A lot of people don't like *any* nuclear power plants, and with some reason—the general idea is that utility company Engineers are in a great rush to build the damned things, while nobody really knows enough yet about what their effects might be—but there wasn't any really great outcry about Indian Point. Like all reactors so far, it is a thermal polluter—it raises the temperature of the river at its site by using river water as a coolant and then dumping the warmed water back into the riverbed—but the Hudson is so fouled up anyway that a lot of people have given up on it.

The only problem was that after the plant opened, somebody noticed a bunch of crows around a dump near the new power station—more crows by far than usual. When this kept up for a while, somebody got in touch with the Long Island League of Saltwater Sportsmen, a group which—in marked contrast to a lot of sportsmen's groups—happened to retain a consulting biologist. His name is Dominick Pirone, and he went to take a look at the dump.

Although Con Ed didn't want him to look, he managed to see bulldozers at work, shoving dead bass into twelve-foot-high piles, where lime was being dumped on them to hasten their decomposition. And he saw a line of trucks, bringing new loads of dead bass. Pirone decided to follow the trucks, and found himself at Indian Point.

"I saw and smelled," he said, "some 10,000 dead and dying fish under the dock."

The seven-degree rise in the temperature of the water, after the plant has sucked in Hudson River water and spat it back out again, is enough to attract spawning bass. They were trapped under the dock and ultimately suffocated. Intake pipes sucked them up into wire baskets, the baskets were dumped into the trucks—and before Con Ed, under pressure, put up a fine-mesh screen around the dock, 2,000,000 bass had been killed.

Con Ed did its best to kill the story. New York's Representative Richard Ottinger (who would probably have leprosy by now if somebody on the Con Ed staff had the Evil Eye) described it this way:

> The story of the Indian Point fish kill is strangely obscure. There are reports of truckloads of fish carted away secretly; fish graveyards limed to hasten the destruction of evidence and guarded by Burns detectives to prevent witnesses' access to see the size of the kill. There are stories of pictures suppressed by State officials and State employees pressured into silence.

That was before Pirone's visit blew it wide open. I've seen some of those once-suppressed pictures and talked to a couple of those state employees. And I've also learned that fine-mesh screens may prevent spectacular fish kills—but they don't necessarily protect fish, as will become clear when we get to Storm King.

Before that, though, let us turn to Wesley Marx, from whose book the Indian Point account is largely taken. He uses it, as it must be used, as an example of the need for an ecological sense, an ecological conscience—not as a moral imperative but as a practical need in an increasingly hungry world where every salmon, bass and shad may be of literally vital importance:

> To be fully effective any program designed to preserve anadromous fish [salt-water fish who go up fresh-water streams to spawn] from extinction in the twentieth century must encompass the entire environment of the fish, from the international waters of the open ocean to the headwaters of mountain streams. Every human activity along this vast range, from fishing to waste disposal to water storage, must meet the breathing, feeding, and spawning requirements of the fish. Wherever these requirements clash with human activity, it would have to cease or a duplicate environment for the fish created [*sic*]. A program of anadromous conservation thus runs up against every conceivable political and economic obstacle in modern life.

But it's been done on the Columbia, and it could be done on the Hudson. The biggest "political and economic obstacle" is Con Ed.

On a foggy January afternoon, I drove slowly up Route 9D in New York, a winding, occasionally dangerous road that follows the east bank of the Hudson. To my immediate left, railroad yards, dumps and frequent junk piles occupied the foreground; in the middle distance the brown and dirty but still-impressive river went past in the other direction, as much as three and a half miles wide at Haverstraw Bay; beyond, in the distance, the false fronts of the Palisades, looming and impressive even though I knew that their backs have been quarried away, looked somberly down.

I did not think about Henry Hudson and the tiny *Half Moon*, probing 350 years ago for a passageway to the western sea, nor about Benson Lossing a century ago, paddling and sketching and taking notes for his still-beautiful book on the river. Instead, I found myself thinking about the desperate people of Manhattan, where for a time I lived, and whose citizens seemed to share an almost frantic need to get out, to get away, at every opportunity, while at the same time they cannot or will not give up the fantastic complexity and excitement of that filthy, exciting, lonely, exhilarating city. I thought of the power blackout, through which I had lived, and how amiably New Yorkers had borne it—and of how dramatically it demonstrated the city's total dependence on electricity and on the Consolidated Edison Company.

Where a tunnel pokes through Breakneck Ridge, I stopped, left the car in the parking lot of the deserted Breakneck Lodge, and climbed down the bank to walk across the New York Central tracks and stand on the shore. Across the Hudson—narrow and hurrying at this point—was Storm King Mountain.

The ridge behind me was, and is, 295 feet higher than Storm King's 1,340 feet; I remember thinking irrelevantly that the highest point in my native hilly San Francisco is less than 1,000 feet above sea level. But Breakneck Ridge is simply there. Storm King rises directly and abruptly from the rushing river, and in its presence, even from the other side of the river, I felt that sense of helpless puniness that men are apt to feel in the presence of nature at its most impressive.

The Engineers of the Consolidated Edison Company feel no such puniness. Like all of the Engineers, they know of mountains and rivers only that they are there to be used, changed, managed. Storm King, they have decided, is the ideal place for a pump storage plant.

A pump storage plant is a simple thing, though not in itself an economical one. Its function is to suck up water, and to pump it to a high reservoir. Then, at peak periods when the company needs to deliver a lot of power, the reservoir releases the water to fall back into the river, and the plant uses the falling water to generate electricity.

The drawings for the plant, which was to be on the north slope of Storm King, showed no particular esthetic damage to the site, but then of course they wouldn't. When conservationists protested the plans to build the plant, Con Ed promptly responded, not only that "we need the power" and that taxes and payroll would benefit the town of Cornwall (which rose magnificently to the bait), but that the care which Con Ed would give the site would actually improve its scenic values.

The company also said it would build the town of Cornwall, in which Storm King is located, a riverfront park (since they have to dump all that dirt somewhere, they plan simply to dump it into the river—which says enough about their concern for bass spawning grounds and shoreline ecology—and then to build the park on it)—which the town welcomed, partly because a park built without state or Federal funds could be restricted to local citizens. The town fathers call this being free from "urban pressures." What it really means is that they can keep out not only Negroes and Puerto Ricans from New York City, but even the nearest black citizens, from Newburgh.

By now, however, we have heard all this nonsense about improving the scenery before. A seacoast away, the proposal of PG&E to deface California's beautiful Bodega Head with a nuclear reactor was billed by the company as "PG&E's Atomic Park." In neither of these two cases, as it happens, did the Engineers get away with the peculiar esthetics of their first attempt; public pressure forced PG&E to surrender its reactor plans (temporarily; it has leased the site to the county as a park for $1 a year, but for only five years), and on the Hudson, Con Ed finally, and grumpily, agreed that it could bury its 800-foot-long plant, pump its water up and let it fall through a tunnel, and put the necessary power lines underground instead of allowing them to march through the town of Yorktown on steel towers.

Now—as major industries usually do—they are taking all the credit for what they were forced to do; the "underground plant" and the "hidden tunnel" are trumpeted as evidence of their concern for the site, and they profess to be bewildered about why everybody doesn't immediately start to love them for improving Storm King. They have even taken, as New Yorkers will have noticed, to calling it the "Cornwall project" instead of the "Storm King project," a device known to anthropologists and metalinguists as word magic.

But while they are talking fast, loud and long about the great underground hidden operation that no one can see, and simultaneously claiming that it will improve the site, they are hoping that nobody will notice a few other things, and they profoundly wish—with understandable sensitivity after what happened at Indian Point—that everybody would stop talking about fish.

Storm King is uniquely beautiful, and it is of unique historical importance; those facts are actually outside the scope of this book, which is primarily concerned with ecosystems and genetic information. If I go on talking about Storm King's uniqueness a little, it is only because Storm King is such a perfect illustration of an industry's explanation that fractures are better than unbroken ankles. Fish, however, are not outside the scope of this book, and anyone who thinks that a few screens (or, in this case, a whole massive network of screens, a lot of them visible) are going to make it all right just doesn't dig fish.

Testimony from the State Conservation Department should, in a rational world, be enough; and according to New York's Joint Legislative Committee on Natural Resources, "The State Conservation Department admitted that it could not project the effect of the Storm King project on fish life."

Con Ed's paid experts can, of course, and have done so; they say it will be all right. At the first Federal Power Commission hearing on the proposal, someone unkindly pointed out that striped bass—beloved of fisherman from Long Island to South Carolina—go up the Hudson to spawn, and that something like 85 percent of them spawn in the Storm King area. That gigantic "straw" through which Con Ed plans to suck up as much as 1,000,000 cubic feet of water *per minute* from the Hudson can suck up fish and larvae and fish eggs just as easily—and while everybody now knows about the 2,000,000 bass killed at Indian Point, nobody much mentions eggs or larvae.

The FPC ruled once that such warnings were "untimely," and that "the project will not adversely affect the fish resources of the Hudson provided adequate protective facilities are installed." You can't argue with that. The project won't hurt anything provided you build it so that it won't hurt anything. The problem is that you can't do that. An Interior Department spokesman, James McBroom, told a Congressional committee flatly that "practical means of protection of eggs and larvae stages have yet to be devised." Screens just won't do it.

Screens, however, are what Con Ed proposes—and for those who care what all this looks like, the screens include a permanent, but movable, crane to lift them out of the water for cleaning at periodic intervals. What will be lifted out of the water, it appears, are 96 pieces of screen, each 9 feet wide and more than 40 feet deep. They'll hang out there somewhere to mark the site of Washday-on-the-Hudson.

The FPC once granted Con Ed the permit to build this monstrosity, but—happily for the bass and the Hudson—a group called the Scenic Hudson Preservation Conference had come into being, and had hired as executive director a hard-talking, hard-fighting publicist named Rod Vandivert. With the aid of a number of other people, notably attorney David Sive, Vandivert took the FPC to court, and won from the United States Court of Appeals a historic decision:

> In this case, as in many others, the Commission has claimed to be the representative of the public interest. This role does not permit it to act as an umpire blandly calling balls and strikes for adversaries appearing before it; the right of the public must receive active and affirmative protection at the hands of the Commission.

The court ruled that the FPC must take scenic, historic and recreational values into account; that it has to care about the fish; that it has to listen to testimony—such as that offered by former New York City engineer Alexander Lurkis—about alternative methods of providing the power if New York City needs it:

> Especially in a case of this type, where public interest and concern is so great, the Commission's refusal to receive the Lurkis testimony, as well as proffered information on fish protection devices and underground transmission facilities, exhibits a disregard of the statute and of judicial mandates instructing the Commission to probe all feasible alternatives.

Finally, the opinion, written by Judge Paul R. Hays, broke a little new ground in giving explicit directions to the Commission:

> The Commission's renewed proceedings must include as a basic concern the preservation of natural beauty and of national historic shrines, keeping in mind that, in our affluent society, the cost of a project is but one of several factors to be considered.

Nothing could be better calculated to drive the Engineers out of their minds. When the FPC began new hearings in response to the court's order, for instance, *Electrical World*, an industry magazine, said that the "Cornwall project" (they don't call it "Storm King" either, apparently by coincidence) was "tediously being drawn through new but repetitious hearings," and sneered at Con Ed witnesses' being "asked, more or less, to define 'beauty'—in 25 words or less."

To make matters even worse for the Engineers, the Supreme Court came down in early 1967 with what is known as the "High Mountain Sheep decision," concerning a private power company's proposed dam on the Snake River. Secretary of the Interior Stewart Udall had asked the FPC to turn down the application to build the dam, arguing that it might damage fish and wildlife. The FPC said no, and Udall went to court—all the way to the United States Supreme Court. A 7–2 decision said that the test of a hydroelectric project is "whether the project will be in the public interest," and that you determine the public interest by considering

> future power demand and supply, alternate sources of power, the public interest in preserving reaches of wild rivers and wilderness areas, the preservation of anadromous fish for commercial and recreational purposes and the protection of wildlife.

The court also noted, importantly for our purposes:

> The importance of salmon and steelhead in our outdoor life as well as in commerce is so important that there certainly comes a time when their destruction might necessitate a halt in the so-called "improvement" or "development" of waterways.

Striped bass, too, are anadromous fish, and the same argument applies.

Although the FPC doesn't really like all this—it has long since passed from *regulating* public utilities to accommodating them, and thus can be counted as still, more or less, likely to be on Con Ed's side—it is nevertheless true that Con Ed liked it all even less. To the public, the company insistently advertised that "Cornwall represents conservation at its best" (without saying that anything was being conserved—more word magic), and insisted, in italics, that *"It would not damage the wonderful scenery of the Hudson River; being underground, it would not damage the landscape."*

Of course not. Except that—well, for one thing, there's the tailrace.

The spot at Breakneck Ridge from which I first saw Storm King is not the best place. The gap between Storm King and Breakneck is one of the few places—and by far the most impressive and accessible—at which a mountain river breaks through the Appalachian wall to the sea, and the place from which to see it, by land or on the water, is from the north (or from the top of Storm King itself, but we'll get to that later). Here, though you can't see the vital bass, you can see best the majesty and the beauty of the scene.

On the north face, or slope, of Storm King (Con Ed witnesses sometimes argued that the project isn't "on the mountain" at all, but merely on the slope leading up to it, which is still more word magic) is where construction of the project's tailrace—the channel for the water the plant will use—will gouge away several million cubic feet of the mountain.

The river frontage of Storm King is about 6,000 feet. The tailrace gouge will tear away 560 feet of this, and there will be, in addition, "130 feet of supporting concrete abutment at the east end of the bridge and 125 to 150 feet at the west end." In other words, something like 800 feet of the shoreline of Storm King is to be torn away; in place of 560 feet of that shoreline, there will be a hole.

This gouge would go back into the mountain for from 200 to 280 feet—which means, in turn, that the cutting and bulldozing will make a hole back to where the elevation is 70 feet above the river. This is the "underground" project that "would not damage the landscape."

Con Ed has promised replanting, and one of the landscape architects who testified for them—another living demonstration of the "engineering mentality" in another profession—said that "the concrete bridges at the mouth of the tailrace will be treated so that they will blend in with the background." Camouflage paint, no doubt.

But that isn't all. There is a big "recreational plan" in the works, too, to be built on the mountain itself (no argument about that, this time). Lest I be accused of unfairness, let's let some of this be described by one of the men who designed it and who testified for Con Ed:

> The visitors' information building will contain a reception room, display area, a seminar room and a small auditorium, together with necessary sanitary facilities. It will also contain an elevator which will provide a means of access to an observation room located far below in the power plant itself. . . .
>
> An access road from Route 218 to the general site will be constructed. Thirty individual picnic units each consisting of a table, benches and fireplace will be built, together with a group picnic shelter with sanitary facilities, fireplace and terraces. A parking lot with space for 40 cars will be provided at the site of the shelter. Along the hard surface entrance road of approximately one mile, a number of two car turnoffs will be constructed. Additionally, lawn areas will be available where group games may be played and outdoor concerts held.

None of which will mar the landscape, of course.

The visitors' information center (according to a Con Ed exhibit) will be 150 feet long and 37 feet high, and "several appropriately landscaped observation terraces will be placed adjacent to the building to afford panoramic views of the Hudson River and the surrounding countryside." In addition, there'll be another section of parking lot; together, the two sections will have a capacity of 120 automobiles and three buses.

But it's okay. They're going to plant something around all of it. One of the landscape Engineers said that "the planting around the Visitors' Information Center will make the Center virtually invisible from the river and from the opposite shore"—but it will have panoramic views. Presumably through peepholes.

Actually, Con Ed's own exhibits contradict the invisibility claim, and the same landscaper admitted that between October 15 and May 15, the planting will *not*, in fact, make the center invisible, virtually or otherwise. One would suppose, in addition, that it is during the remaining months, between May and October, that we might expect to find the most people using those 31 invisible fireplaces. But the mountain will be prettier, anyway—even though it is already impressive enough to have moved James Fenimore Cooper to thoughts of Creation and to have inspired Washington Irving.

And then there's Cornwall's lily-white park. It will be 57 acres in size; there will be about a mile of roadway with "appropriate parking space," 10 acres of open lawn, 30 picnic units (each with a fireplace), some paved areas, two group picnic shelters and sanitary facilities. None of which anybody even claims will be invisible (another 30 invisible fireplaces would apparently be too much even for Con Ed's word magicians). It will just be prettier.

It will, too. The base of Storm King today, as the Engineers are fond of pointing out, is not all that pretty up close. It is piled up with junk, in fact. But you can pick up junk without dumping millions of tons of mountain into the river and making a phony shoreline. "The disposal of fill to base up the park," testified Charles W. Eliot II—a fairly well-known landscape architect in his own right—"will replace two long sections of *natural* shoreline with practically straight line rip-rap or rock wall [his emphasis]," and he continued:

> . . . the artificiality of the new rip-rap shore, together with what is typical of the usual city park, will further emphasize the interference of the whole project with the natural conditions along the river and Storm King.

As a glance at the exhibits shows, the park will also wreck the curve of the bank out of which the Storm King promontory now so dramatically thrusts. And then there's the reservoir. Con Ed's ads don't even *mention* the reservoir, and I don't blame them.

The reservoir is in the valley between Mount Misery and White Horse Mountain, in part of the Black Rock Forest. It is necessary because, of course, when they suck up 1,000,000 cubic feet of the Hudson every minute or so, they have to put it somewhere. They've decided to cover a treasured forest that is used as a study area by Harvard's botany department and includes trails maintained by the New York-New Jersey Trail Conference (although Con Ed's experts say it's full of rattlesnakes).

Again, let's let a Con Ed witness describe this additional piece of rape-as-therapy:

> The project reservoir will be formed by the construction of five dikes of varying sizes in saddles between ridges which presently form a natural basin in the Hudson Highlands in the area of Mount Misery and White Horse Mountain, about 10,000 feet southwest of the underground [*sic*] power plant. The reservoir will occupy a surface area of approximately 240 acres and, at maximum water elevation of 1,160 feet above mean sea level, will have a storage capacity of 25,000 acre-feet, or approximately eight billion gallons.

That word is not "million." It's "billion."

Dr. Eliot, who was *not* testifying for Storm King (or, rather, he *was* testifying for Storm King; he was not testifying for Con Ed), looked at it, however, from a different angle:

> Standing in the area to be flooded, one can get some idea of the size of the future reservoir by imagining water almost to the top of White Horse Mountain and Mount Misery. The water surface will be wider than the Hudson River between Storm King and Breakneck. This huge, man-made water body, almost literally on top of a mountain, will dwarf the scale of Storm King and Crow's Nest. . . .
>
> The main dam across Black Rock Hollow is shown on the plans as 2250 feet long and over 275 feet high at the deepest point. The western slope is shown as extending 500 feet from the face. The other dams required include about 3000 feet on the northeast parallel with Highway 9W; 750 feet east of Mount Misery on the southeast; and 500 plus feet on the north.

None of this testimony was contradicted at the hearing—in fact, Dr. Eliot took the information from the company's own exhibits. But you haven't read about any of this in Con Ed's full-page ads, either.

Their house landscape Engineer said that they might try to disguise the outsides and tops of those "dikes" (a good public relations firm watches every word), but even he had to admit that you can't disguise the inside mud walls as the water rises and falls. The reservoir, incidentally, will be "not safe for recreation uses and therefore it will be necessary to exclude the public from its use." It'll have a fence around it—presumably a fence disguised as a set of ninepins, or a row of forest rangers, or something.

But the 240-acre mud pile and its ugly mud walls will be visible from about fifteen high points in the part of the area that Con Ed will allow people to go on using—a prospect that doesn't bother the Engineer Con Ed presents as its landscape architect. "Any large lake," he said during the hearings, "is handsomer than a small lake."

And in case fifteen high points aren't enough:

> The Company plans to construct a scenic overlook where none presently exists at a point close to Route 9W. This overlook will provide a panoramic view of the Hudson River in one direction and of the project reservoir in the other. The total acreage involved in development is approximately 36 acres. . . . A shelter building with toilets and 24 picnic units with appropriate facilities [fireplaces?] will be constructed for both the overlook and picnic areas.

All invisible, of course. Con Ed's critics insist that, from the company's own plans, the geography of those who planned that overlook is as bad as their esthetics. But there isn't any point in going on with this. There remains only the question: Do we need the power, and if so, what could Con Ed do instead?

At one point, somewhere back in the proceedings, Con Ed gave the cost of Storm King project as $162 million. By the figures of their own expert, the least it can cost them to put the transmission lines underground is $44.5 million. But a company that can put a 70-by-560-foot hole in the side of a mountain and call it "invisible" has little trouble confusing a lay journalist with figures. Because the court ordered the FPC to consider alternatives, the FPC considered alternatives—quite a few of them. And since the court made it quite clear that "it's cheaper" is not, by itself, sufficient excuse for chewing up bass eggs and mountains, a lot of attention was paid to figures.

The minimum differential between Storm King and the next cheapest alternative to provide the same power (I'm assuming that the power is needed, since I haven't really investigated *those* figures) is, according to Con Ed witness Walter Fisk, just under $79 million over twenty years. Con Ed vice president M. L. Waring seems to have gotten this up to $119 million later on. But some of this wasn't too clear, because the Rev. William Hogan, a Fordham economist who was called to bolster Con Ed's position, used the $79-million figure—in fact and to be fair, he spoke of "a cost differential . . . ranging from $78,692,000 to as high as $362,886,000"—but on the same page of testimony he said:

> The alternative closest to the Cornwall project [they call it that at Fordham, too, apparently] in aggregate costs as indicated by Exhibit 257 was the alternative numbered 4A. . . . This alternate . . . would cost . . . $56 million more than the proposed plant at Cornwall.

Well, maybe that's not important. What I really wanted to be sure you don't miss is Father Hogan's reasons why the plant should be built on and in Storm King rather than somewhere else—because you may not realize the far-reaching implications of all this. You see, there are "social implications as well as economic ones involved here." For one thing, 22 percent of the population of New York City is either black or Puerto Rican, and a lot of those blacks and Puerto Ricans are underprivileged. A more expensive alternative to Storm King would—or at least might—mean higher utility rates for these underprivileged people, whereas Storm King might keep their electricity bills down. "Reasonable utility rates," Father Hogan went on, "would encourage the purchase of . . . appliances." These "appliances" would very likely include air conditioning. Air conditioning would give these underprivileged black and Puerto Rican New Yorkers "the prospect of a fuller life," and that would mean that "the threat of hot summers with all their attending implications could be avoided."

In other words, if the FPC will kindly let avuncular old Con Ed build its public service invisible facility at Storm King, it will stop riots in Harlem.

Still granting Con Ed's relative-cost figures to be correct, it turns out, on examination, that even if Con Ed were to lower everybody's electricity bill to match the difference in cost between Storm King and the next cheapest alternative, that underprivileged black family in Harlem will save somewhere between a dime and four bits a year. For fifty cents a year they're all going to buy air conditioners in Harlem, and all the trouble will go away.

There is one further little economic problem here. The alternative in question involves construction, not on Storm King but in New York City itself. That, in turn, means that Con Ed would pay taxes, not to Orange County (where the lily-white Cornwall park would be), but to

New York City. And another economist testified that "$83,000,000 more in local property taxes would be paid to New York City with the alternative" than with Storm King.

You can't buy an air conditioner for four bits a week, but New York City could, if it would, do something about the troubles of the people in Harlem with another 83 million bucks.

Con Ed's concern for being sure that everyone has a broken ankle finally shows up in the rather dry-sounding testimony of Professor Reynold Sachs, a Columbia economist:

> Since, however, estimated costs and taxes extending twenty years into the future necessarily involve forecasts subject to error, the present values described above . . . are not statistically distinguishable. It is accepted scientific practice not to distinguish summary data whose differences are less than the probable errors of measurement inherent in the underlying statistics. In this statistical sense, therefore, and taking account of total costs and benefits attendant upon either proposal, and *using the Applicant's own cost figures*, I can find no statistical basis for distinguishing between the Project and the Alternative [emphasis added].

In other words, keep your eye on the pea.

The new hearings ended on May 23, 1967, a bookshelf full of briefs have been filed, and the FPC staff is still in there pitching for Con Ed. For instance, at page 13 in the staff brief, it says, "Despite the fact that the powerhouse and appurtenant works would no longer be seen at all, opposition still persisted. . . ." Even the Con Ed witnesses never made that claim. The FPC staff, instead of reading the Con Ed testimony and accepting it. apparently devoted itself to reading the Con Ed ads in *The New York Times.*

Following closely the arguments in the staff brief, the Hearing Examiner for the FPC recommended, in mid-1968, that the plant be built. In October briefs were filed by both sides. No matter how the Commission finally rules, the whole thing will probably go to court, and the argument may go on for years—because Con Ed is clearly going to go on insisting that concepts like beauty, history and ecology have no place in the construction of a technological society.

After a while you have to get off the subject. You can take just so much solemn testimony and public relations balderdash about holes improving mountains and cartilage being stronger than bone. Anger starts to set in, and you want to talk about something else. You can't, though. Everywhere you look, everywhere where there is something that really ought to be saved, there is someone to assure us that we would be better off if the ankle were broken—and other industries are almost as good at it as utilities.

A few years ago, Governor (now Senator) Gaylord Nelson of Wisconsin discovered that 32 percent of the shallow wells in Wisconsin were polluted by detergents. The resulting investigation showed that the pollution was present in wells all over the country.

The detergents—made from a petroleum-based chemical, alkyl benzene sulfonate—resisted attack by the bacteria in sewage and waste water, and simply wouldn't break down (in scientific language, it was non-biodegradable). They poured out of sewage disposal plants in their original form, and poured down rivers and into lakes as mountains of foam. In March, 1965, the Milwaukee River near West Bend, Wisconsin, was covered with a mountain of detergent foam 40 feet high. A community-owned well in Yuba City, California, had to be abandoned because of ABS pollution. The United States Fish and Wildlife Service reported that ABS detergents "were toxic to eggs and larvae of clams and oysters at concentrations of 0.6 parts per million or less." And of course, as water is used and reused, detergent concentrations become higher.

The detergent industry, however, insisted loudly and publicly that there was nothing to worry about. In fact, they insisted on recommending fractures to all their patients. The board of directors of the Soap and Detergent Association met in April, 1963, and E. Scott Pattison issued a press release following their meeting, arguing that detergent foam in wastes "serves in many cases to warn that other less feasible [*sic*] and more dangerous pollutants are present."

Six months later, the Monsanto Company published an article stating that detergent pollution had never, ever hurt anybody, and suggesting that ABS "may be valuable as a tracer which indicates that more dangerous sewage ingredients, including disease organisms, may be reaching the water supply." They did not go quite so far as to suggest that drinking All would prevent or cure cholera, but they did argue that legislative remedy was a ridiculous idea. Scientifically, they argued against a bill introduced by Senators Nelson and Maurine Neuberger of Oregon, which set a deadline of June 30, 1965, by which the manufacturers had to make the detergents biodegradable or quit making them at all. That was asking for the impossible, the manufacturers insisted; the cost would be millions, and the legislation would impede scientific progress.

They did it, though—just as the auto makers met safety standards when they were forced to, although that was impossible too. By June of 1965, the Soap and Detergent Association's members had "voluntarily" changed over their detergent manufacture so that today's detergents are based on linear alkylate sulfonate—a "soft," or biodegradable, compound.

Even LAS is not as good as it might be, and further refinement of detergents is under way. But the point is that left to themselves, the soap people, like the auto people and the public utilities people, are among the leading advocates of that great Engineer's concept immortalized by Joseph Heller as "catch-22": They have a right to do anything we can't stop them from doing.

The broken-ankles-are-better philosophy often comes out—as it did in the speech by lumberman Buchanan—as the philosophy of "maximum use" (or, sometimes, "multiple use"). Industrial polluters of water, in the words of the Conservation Foundation's Russell Train, put it this way:

> A stream has a natural capacity to assimilate waste. This assimilative capacity (which depends upon the availability of free oxygen in the stream) is a natural resource. Conservation means wise multiple use of natural resources. Therefore, it is 'true conservation' to use the assimilative capacity.

Sound like "Cornwall represents conservation at its best"?

Having quoted it, Train proceeds to tear it apart, but it hardly seems necessary; the sophism is self-evident. It is similar to the argument that underground transmission lines shouldn't be used because they raise the cost of power, or that pollution abatement devices can't be installed because they raise the price of manufactured goods. "Following this approach," Train said in a different speech, "child labor would never have been abolished."

But there is no stopping them. *The Washington Star* noted early in 1967 that billboard companies were still objecting to a proposed control law because signs would have to be kept 2,000 feet away from on- and off-ramps, and that in commercial and industrial areas signs would have to be spaced at least 500 feet apart. At that distance, you would pass ten signs a minute if you were driving at 60 miles an hour—and that was a billboard "control" bill, unacceptable to the industry.

Over a hundred years ago, Henry David Thoreau gazed with sick horror on the work of Engineers on the Concord River, and wrote: "Poor shad, where is thy redress . . . who hears the fishes when they cry?" Today, approximately a hundred miles away, the Nature Conservancy's Dr. Richard Goodwin reports that "the Yankee atomic [power] plant on the lower Connecticut River is very likely to destroy the highly productive shad run in the river. What provisions have been made to avoid this? A study—to be made only through the industry, whose profits are involved, to determine whether damage is done. Only then will cooling towers be installed. By this time, it is not at all unlikely that remedial action will be too late."

The Engineers of private profit on the Connecticut use the same argument as the Engineers of the Army on canal C-111 in the Everglades: we'll do it first and see what happens. Neither ecology nor beauty enters into their calculations. In 1966, a House sub-committee held lengthy hearings on the general subject of pollution. Through the hundreds of pages of testimony one theme recurs again and again: Industries reject or resist pollution abatement devices, because they are an investment without an immediate return.

In the meantime, the profitmakers and their Engineers march on: electric plants on the upper Savannah and the beautiful St. Croix, pulp mills pouring their garbage into the Great Lakes, lumbermen nibbling at the forests and waiting for the regulators to look the other way. Even the Adirondack Forest Preserve . . . has been quietly, but legally, invaded. During World War II, because of the emergency, New York allowed the Tahawus Railroad to be built, to haul out precious titanium ore from a mine operated by the National Lead Company. Militantly, the state insisted that the agreement be written to cover only the time of national war emergency, though they allowed a fifteen-year margin at the end to allow for making other arrangements, ripping out tracks, etc. About three years ago, without a word being said in public, the state of New York (in violation of its own constitution) quietly extended the agreement for 100 years. The Federal Government was just involved enough so that strict legalities were observed and the state's constitution, with its "forever wild" clause, could be overridden. Where there is a buck, someone will do his best to find a way; and if he's caught, he'll find a way to tell you that it's all, really, good for you.

The Kennecott Copper Corporation—aside from a spokesman's occasional and nonsensical reference to the war in Vietnam—has not yet told us how its open-pit mine will improve the scenic value of the Cascades or make the steelhead in the rivers healthier. But pollutants can be tracers for disease organisms; a Con Ed pump storage plant can improve the impressive majesty of the Hudson's gorge at Storm King; PG&E can propose an "atomic park," and I have just heard a radio commercial for "San Mateo's *beautiful* Hillsdale shopping center, where there are acres of free parking just a few steps away." Certainly we can expect some word magic from the copper kings any day now about how we'd all be better off with broken ankles.

Con Edison Plant Shut Three Days to Stop Fish-Kill

Peter Millones

The Consolidated Edison Company closed its Indian Point One nuclear plant at Buchanan, N.Y., for three days last month because "a very substantial number" of fish were killed when they were drawn from the Hudson River into conduits that supply water to the plant.

Just how many fish were killed has not yet been determined, according to the State Conservation Department, which is investigating. A Con Edison official said that the company viewed the problem as "embarrassing and serious," but that the "dimensions of the kill do not appear to be massive." At the same time, he said, the company is not relying on the electrical-generating capacity of the plant in case it needs to be shut again.

The killing of striped bass, white perch and other fish was not related to radiation or thermal pollution, but instead was similar—though apparently less extensive—to a fish kill at the same Indian Point plant in 1963. The fish then and now, for some unexplained reason, are occasionally attracted to the large concrete conduits that carry 350,000 gallons of river water a minute into the plant to cool the superheated plant condensers.

Inside the conduits are straining screens, designed to filter debris from the river. The fish are caught on the screen and killed, although some small fish were said to have gone farther into the system through an opening believed to be at the bottom of one screen. Some fish also entered deep into the conduit according to an official in the State Conservation Department, because in January's unusually cold weather,

From the *New York Times* (February 3, 1970).

ice formed on the screens and the screens were raised by workmen to eliminate ice and allow water to pass through. Con Edison sent a diver down in January to examine the screens. This led to the suspicion of erosion beneath a screen.

In 1963, conservationists and Representative Richard L. Ottinger, Democrat of Westchester, charged that the screen problem had killed "hundreds of thousands to millions" of fish. Con Edison denied that the number of dead fish was that large.

Until last December, the company thought that its new screens had solved the problem. But on January 19th the nuclear plant was closed and it was re-opened on January 22nd. The problem had not been solved, but it was impossible to tell if the changes in the screens were effective when the plant was not drawing water. Some conservationists have opposed power plants that draw water from rivers because they contend that no screens can effectively keep out eggs and larval-sized fish.

William Bentley, assistant director of the Fish and Wildlife Bureau, characterized the number of fish killed as "very substantial" and said that state officials were at the plant daily now working with Con Edison engineers on a possible solution. The kill is likely to affect the long controversy surrounding Con Edison's desire to build farther up the Hudson at Cornwall a pumped-storage plant that would also draw great quantities of water from the river.

The company has contended that opposition to its plans by conservationists was a major factor in the shortage of electric power here last summer and projected for the coming summer.

PART IV

CITIES

City Life

When I am in a great city, I know that I despair.
I know there is no hope for us, death waits, it is useless
to care.
For oh the poor people, that are flesh of my flesh,
I, that am flesh of their flesh,
when I see the iron hooked into their faces
their poor, their fearful faces
I scream in my soul, for I know I cannot
take the iron hooks out of their faces, that make them so
drawn,
nor cut the invisible wires of steel that pull them
back and forth, to work,
back and forth to work,
like fearful and corpus-like fishes hooked and being played
by some malignant fisherman on an unseen shore
where he does not choose to land them yet, hooked fishes
of the factory world.

D. H. Lawrence

From *The Complete Poems of D. H. Lawrence*, Vol. II, edited by Vivian de Sola Pinto and F. Warren Roberts (New York, 1964).

Future Times

Anatole France

1

The houses were never high enough to satisfy them; they kept on making them still higher and built them of thirty or forty storeys with offices, shops, banks, societies one above another; they dug cellars and tunnels ever deeper downwards.

Fifteen millions of men laboured in a giant town by the light of beacons which shed forth their glare both day and night. No light of heaven pierced through the smoke of the factories with which the town was girt, but sometimes the red disk of a rayless sun might be seen riding in the black firmament through which iron bridges ploughed their way, and from which there descended a continual shower of soot and cinders. It was the most industrial of all the cities in the world and the richest. Its organisation seemed perfect. None of the ancient aristocratic or democratic forms remained; everything was subordinated to the interests of the trusts. This environment gave rise to what anthropologists called the multi-millionaire type. The men of this type were at once energetic and frail, capable of great activity in forming mental combinations and of prolonged labour in offices, but men whose nervous irritability suffered from hereditary troubles which increased as time went on.

Like all true aristocrats, like the patricians of republican Rome or the squires of old England, these powerful men affected a great severity in their habits and customs. They were the ascetics of wealth. At the meetings of the trusts an observer would have noticed their smooth and puffy faces, their lantern cheeks, their sunken eyes and wrinkled brows. With bodies more withered, complexions yellower, lips drier, and eyes

From *Penguin Island*, by Anatole France (New York and London, 1909), pp. 326-345.

filled with a more burning fanaticism than those of the old Spanish monks, these multi-millionaires gave themselves up with inextinguishable ardour to the austerities of banking and industry. Several, denying themselves all happiness, all pleasure, and all rest, spent their miserable lives in rooms without light or air, furnished only with electrical apparatus, living on eggs and milk, and sleeping on camp beds. By doing nothing except pressing nickel buttons with their fingers, these mystics heaped up riches of which they never even saw the signs, and acquired the vain possibility of gratifying desires that they never experienced.

The worship of wealth had its martyrs. One of these multi-millionaires, the famous Samuel Box, preferred to die rather than surrender the smallest atom of his property. One of his workmen, the victim of an accident while at work, being refused any indemnity by his employer, obtained a verdict in the courts, but repelled by innumerable obstacles of procedure, he fell into the direst poverty. Being thus reduced to despair, he succeeded by dint of cunning and audacity in confronting his employer with a loaded revolver in his hand, and threatened to blow out his brains if he did not give him some assistance. Samuel Box gave nothing, and let himself be killed for the sake of principle.

Examples that come from high quarters are followed. Those who possessed some small capital (and they were necessarily the greater number), affected the ideas and habits of the multi-millionaries, in order that they might be classed among them. All passions which injured the increase or the preservation of wealth, were regarded as dishonourable; neither indolence, nor idleness, nor the taste for disinterested study, nor love of the arts, nor, above all, extravagance, was ever forgiven; pity was condemned as a dangerous weakness. Whilst every inclination to licentiousness excited public reprobation, the violent and brutal satisfaction of an appetite was, on the contrary, excused; violence, in truth, was regarded as less injurious to morality, since it manifested a form of social energy. The State was firmly based on two great public virtues: respect for the rich and contempt for the poor. Feeble spirits who were still moved by human suffering had no other resource than to take refuge in a hypocrisy which it was impossible to blame, since it contributed to the maintenance of order and the solidity of institutions.

Thus, among the rich, all were devoted to the social order, or seemed to be so; all gave good examples, if all did not follow them. Some felt the severity of their position cruelly; but they endured it either from pride or from duty. Some attempted, in secret and by subterfuge, to escape from it for a moment. One of these, Edward Martin, the President of the Steel Trust, sometimes dressed himself as a poor man, went forth to beg his bread, and allowed himself to be jostled by the passers-by. One day, as he asked alms on a bridge, he engaged in a quarrel with a real beggar, and filled with a fury of envy, he strangled him.

As they devoted their whole intelligence to business, they sought no intellectual pleasures. The theatre, which had formerly been very flourishing among them, was now reduced to pantomimes and comic dances. Even the pieces in which women acted were given up; the taste for pretty forms and brilliant toilettes had been lost; the somersaults of clowns and the music of negroes were preferred above them, and what roused enthusiasm was the sight of women upon the stage whose necks were bedizened with diamonds, or processions carrying golden bars in triumph. Ladies of wealth were as much compelled as the men to lead a respectable life. According to a tendency common to all civilizations, public feeling set them up as symbols; they were, by their austere magnificence, to represent both the splendour of wealth and its intangibility. The old habits of gallantry had been reformed, but fashionable lovers were now secretly replaced by muscular labourers or stray grooms. Nevertheless, scandals were rare, a foreign journey concealed nearly all of them, and the Princesses of the Trusts remained objects of universal esteem.

The rich formed only a small minority, but their collaborators, who composed the entire people, had been completely won over or completely subjugated by them. They formed two classes, the agents of commerce or banking, and workers in the factories. The former contributed an immense amount of work and received large salaries. Some of them succeeded in founding establishments of their own; for in the constant increase of the public wealth the more intelligent and audacious could hope for anything. Doubtless it would have been possible to find a certain number of discontented and rebellious persons among the immense crowd of engineers and accountants, but this powerful society had imprinted its firm discipline even on the minds of its opponents. The very anarchists were laborious and regular.

As for the workmen who toiled in the factories that surrounded the town, their decadence, both physical and moral, was terrible; they were examples of the type of poverty as it is set forth by anthropology. Although the development among them of certain muscles, due to the particular nature of their work, might give a false idea of their strength, they presented sure signs of morbid debility. Of low stature, with small heads and narrow chests, they were further distinguished from the comfortable classes by a multitude of physiological anomalies, and, in particular, by a common want of symmetry between the head and the limbs. And they were destined to a gradual and continuous degeneration, for the State made soldiers of the more robust among them, and the health of these did not long withstand the brothels and the drink-shops that sprang up around their barracks. The proletarians became more and more feeble in mind. The continued weakening of their intellectual faculties was not entirely due to their manner of life; it resulted also from a methodical selection carried out by the employers. The latter, fearing that workmen of too great ability might be inclined to put forward legitimate demands, took care to eliminate them by every possible means, and preferred to engage ignorant and stupid labourers, who were incapable of defending their rights, but were yet intelligent enough to perform their toil, which highly perfected machines rendered extremely simple. Thus the proletarians were unable to do anything to improve their lot. With difficulty did they succeed by means of strikes in maintaining the rate of their wages. Even this means began to fail them. The alternations of production inherent in the capitalist system caused such cessations of work that, in several branches of industry, as soon as a strike was declared, the accumulation of products allowed the employers to dispense with the strikers. In a word, these miserable employees were plunged in a gloomy apathy that nothing enlightened and nothing exasperated. They were necessary instruments for the social order and well adapted to their purpose.

Upon the whole, this social order seemed the most firmly established that had yet been seen, at least among mankind, for that of bees and ants is incomparably more stable. Nothing could foreshadow the ruin of a system founded on what is strongest in human nature, pride and cupidity. However, keen observers discovered several grounds for uneasiness. The most certain, although the least apparent, were of an economic order, and consisted in the continually increasing amount of over-production, which entailed long and cruel interruptions of labour, though these were, it is true, utilized by the manufacturers as a means of breaking the power of the workmen, by facing them with the prospect of a lock-out. A more obvious peril resulted from the physiological state of almost the entire population. "The health of the poor is what it must be," said the experts in hygiene, "but that of the rich leaves much to be desired." It was not difficult to find the causes of this. The supply of oxygen necessary for life was insufficient in the city, and men breathed in an artificial air. The food trusts, by means of the most daring chemical syntheses, produced artificial wines, meat, milk, fruit, and vegetables, and the diet thus imposed gave rise to stomach and brain troubles. The multi-millionaires were bald at the age of eighteen; some showed from time to time a dangerous weakness of mind. Over-strung and enfeebled, they gave enormous sums to ignorant charlatans; and it was a common thing for some trumpery bath-attendant or other who turned healer or prophet, to make a rapid fortune by the practice of medicine or theology. The number of lunatics increased continually; suicides multiplied in the world of wealth, and many of them were accompanied by atrocious and extraordinary circumstances, which bore witness to an unheard of perversion of intelligence and sensibility.

Another fatal symptom created a strong impression upon average minds. Terrible accidents, henceforth periodical and regular, entered into people's calculations, and kept mounting higher and higher in statistical tables. Every day, machines burst into fragments, houses fell down, trains laden with merchandise fell on to the streets, demolishing entire buildings and crushing hundreds of passers-by. Through the ground, honey-combed with tunnels, two or three storeys of work-shops would often crash, engulfing all those who worked in them.

2

In the southwestern district of the city, on an eminence which had preserved its ancient name of Fort Saint-Michel, there stretched a square where some old trees still spread their exhausted arms above the greensward. Landscape gardeners had constructed a cascade, grottos, a torrent, a lake, and an island, on its northern slope. From this side one could see the whole town with its streets, its boulevards, its squares, the multitude of its roofs and domes, its air-passages, and its crowds of men, covered with a veil of silence, and seemingly enchanted by the distance. This square was the healthiest place in the capital; here no smoke obscured the sky, and children were brought here to play. In summer some employees from the neighbouring offices and laboratories used to resort to it for a moment after their luncheons, but they did not disturb its solitude and peace.

It was owing to this custom that, one day in June, about mid-day, a telegraph clerk, Caroline Meslier, came and sat down on a bench at the end of a terrace. In order to refresh her eyes by the sight of a little green, she turned her back to the town. Dark, with brown eyes, robust and placid, Caroline appeared to be from twenty-five to twenty-eight years of age. Almost immediately, a clerk in the Electricity Trust, George Clair, took his place beside her. Fair, thin, and supple, he had features of a feminine delicacy; he was scarcely older than she, and looked still younger. As they met almost every day in this place, a comradeship had sprung up between them, and they enjoyed chatting together. But their conversation had never been tender, affectionate, or even intimate. Caroline, although it had happened to her in the past to repent of her confidence, might perhaps have been less reserved had not George Clair always shown himself extremely restrained in his expressions and behaviour. He always gave a purely intellectual character to the conversation, keeping it within the realm of general ideas, and, moreover, expressing himself on all subjects with the greatest freedom.

He spoke frequently of the organization of society, and the conditions of labour.

"Wealth," said he, "is one of the means of living happily; but people have made it the sole end of existence."

And this state of things seemed monstrous to both of them.

They returned continually to various scientific subjects with which they were both familiar.

On that day they discussed the evolution of chemistry.

"From the moment," said Clair, "that radium was seen to be transformed into helium, people ceased to affirm the immutability of simple bodies; in this way all those old laws about simple relations and about the indestructibility of matter were abolished."

"However," said she, "chemical laws exist."

For, being a woman, she had need of belief.

He resumed carelessly:

"Now that we can procure radium in sufficient quantities, science possesses incomparable means of analysis; even at present we get glimpses, within what are called simple bodies, of extremely diversified complex ones, and we discover energies in matter which seem to increase even by reason of its tenuity."

As they talked, they threw bits of bread to the birds, and some children played around them.

Passing from one subject to another:

"This hill, in the quaternary epoch," said Clair, "was inhabited by wild horses. Last year, as they were tunnelling for the water mains, they found a layer of the bones of primeval horses."

She was anxious to know whether, at that distant epoch, man had yet appeared.

He told her that man used to hunt the primeval horse long before he tried to domesticate him.

"Man," he added, "was at first a hunter, then he became a shepherd, a cultivator, a manufacturer . . . and these diverse civilizations succeeded each other at intervals of time that the mind cannot conceive."

He took out his watch.

Caroline asked if it was already time to go back to the office.

He said it was not, that it was scarcely half-past twelve.

A little girl was making mud pies at the foot of their bench; a little boy of seven or eight years was playing in front of them. Whilst his mother was sewing on an adjoining bench, he played all alone at being a run-away horse, and with that power of illusion, of which children are capable, he imagined that he was at the same time the horse, and those who ran after him, and those who fled in terror before him. He kept struggling with himself and shouting: "Stop him, Hi! Hi! This is an awful horse, he has got the bit between his teeth."

Caroline asked the question:

"Do you think that men were happy formerly?"

Her companion answered:

"They suffered less when they were younger. They acted like that little boy: they played; they played at arts, at virtues, at vices, at heroism, at beliefs, at pleasures; they had illusions which entertained them; they made a noise; they amused themselves. But now. . . ."

He interrupted himself, and looked again at his watch.

The child, who was running, struck his foot against the little girl's pail, and fell his full length on the gravel. He remained a moment stretched out motionless, then raised himself up on the palms of his hands. His forehead puckered, his mouth opened, and he burst into tears. His mother ran up, but Caroline had lifted him from the ground and was wiping his eyes and mouth with her handkerchief. The child kept on sobbing and Clair took him in his arms.

"Come, don't cry, my little man! I am going to tell you a story.

"A fisherman once threw his net into the sea and drew out a little, sealed, copper pot, which he opened with his knife. Smoke came out of it, and as it mounted up to the clouds the smoke grew thicker and thicker and became a giant who gave such a terrible yawn that the whole world was blown to dust. . . ."

Clair stopped himself, gave a dry laugh, and handed the child back to his mother. Then he took out his watch again, and kneeling on the bench with his elbows resting on its back he gazed at the town. As far as the eye could reach, the multitude of houses stood out in their tiny immensity.

Caroline turned her eyes in the same direction.

"What splendid weather it is!" said she. "The sun's rays change the smoke on the horizon into gold. The worst thing about civilization is that it deprives one of the light of day."

He did not answer; his looks remained fixed on a place in the town.

After some seconds of silence they saw about half a mile away, in the richer district on the other side of the river, a sort of tragic fog rearing itself upwards. A moment afterwards an explosion was heard even where they were sitting, and an immense tree of smoke mounted towards the pure sky. Little by little the air was filled with an imperceptible murmur caused by the shouts of thousands of men. Cries burst forth quite close to the square.

"What has been blown up?"

The bewilderment was great, for although accidents were common, such a violent explosion as this one had never been seen, and everybody perceived that something terribly strange had happened.

Attempts were made to locate the place of the accident; districts, streets, different buildings, clubs, theatres, and shops were mentioned. Information gradually became more precise and at last the truth was known.

"The Steel Trust has just been blown up."

Clair put his watch back into his pocket.

Caroline looked at him closely and her eyes filled with astonishment.

At last she whispered in his ear:

"Did you know it? Were you expecting it? Was it you . . ."

He answered very calmly:

"That town ought to be destroyed."

She replied in a gentle and thoughtful tone:

"I think so too."

And both of them returned quietly to their work.

3

From that day onward, anarchist attempts followed one another every week without interruption. The victims were numerous, and almost all of them belonged to the poorer classes. These crimes roused public resentment. It was among domestic servants, hotel-keepers, and the employees of such small shops as the Trusts still allowed to exist, that indignation burst forth most vehemently. In popular districts women might be heard demanding unusual punishments for the dynamitards. (They were called by this old name, although it was hardly appropriate to them, since, to these unknown chemists, dynamite was an innocent material only fit to destroy ant-hills, and they considered it mere child's play to explode nitro-glycerine with a cartridge made of fulminate of mercury.) Business ceased suddenly, and those who were least rich were the first to feel the effects. They spoke of doing justice themselves to the anarchists. In the mean time the factory workers remained hostile or indifferent to violent action. They were threatened, as a result of the decline of business, with a likelihood of losing their work, or even a lock-out in all the factories. The Federation of Trade Unions proposed a general strike as the most powerful means of influencing the employers, and the best aid that could be given to the revolutionists, but all the trades with the exception of the gilders refused to cease work.

The police made numerous arrests. Troops summoned from all parts of the National Federation protected the offices of the Trusts, the houses of the multi-millionaires, the public halls, the banks, and the big shops. A fortnight passed without a single explosion, and it was concluded that the dynamitards, in all probability but a handful of persons, perhaps even still fewer, had all been killed or captured, or that they were in hiding, or had taken flight. Confidence returned; it returned at first among the poorer classes. Two or three hundred thousand soldiers, who had been lodged in the most closely populated districts, stimulated trade, and people began to cry out: "Hurrah for the army!"

The rich, who had not been so quick to take alarm, were reassured more slowly. But at the Stock Exchange a group of "bulls" spread optimistic rumours and by a powerful effort put a brake upon the fall in prices. Business improved. Newspapers with big circulations supported the movement. With patriotic eloquence they depicted capital as laughing in its impregnable position at the assaults of a few dastardly criminals, and public wealth maintaining its serene ascendency in spite of the vain threats made against it. They were sincere in their attitude, though at the same time they found it benefited them. Outrages were forgotten or their occurrence denied. On Sundays, at the race-meetings, the stands were adorned by women covered with pearls and diamonds. It was observed with joy that the capitalists had not suffered. Cheers were given for the multi-millionaires in the saddling rooms.

On the following day the Southern Railway Station, the Petroleum Trust, and the huge church built at the expense of Thomas Morcellet were all blown up. Thirty houses were in flames, and the beginning of a fire was discovered at the docks. The firemen showed amazing intrepidity and zeal. They managed their tall fire-escapes with automatic precision, and climbed as high as thirty storeys to rescue the luckless inhabitants from the flames. The soldiers performed their duties with spirit, and were given a double ration of coffee. But these fresh casualties started a panic. Millions of people, who wanted to take their money with them and leave the town at once, crowded the great banking houses. These establishments, after paying out money for three days, closed their doors amid mutterings of a riot. A crowd of fugitives, laden with their baggage, besieged the railway stations and took the town by storm. Many who were anxious to lay in a stock of provisions and take refuge in the cellars, attacked the grocery stores, although they were guarded by soldiers with fixed bayonets. The public authorities displayed energy. Numerous arrests were made and thousands of warrants issued against suspected persons.

During the three weeks that followed no outrage was committed. There was a rumour that bombs had been found in the Opera House, in the cellars of the Town Hall, and beside one of the pillars of the Stock Exchange. But it was soon known that these were boxes of sweets that had been put in those places by practical jokers or lunatics. One of the accused, when questioned by a magistrate, declared that he was the chief author of the explosions, and said that all his accomplices had lost their lives. These confessions were published by the newspapers and helped to reassure public opinion. It was only towards the close of the examination that the magistrates saw they had to deal with a pretender who was in no way connected with any of the crimes.

The experts chosen by the courts discovered nothing that enabled them to determine the engine employed in the work of destruction. According to their conjectures the new explosive emanated from a gas which radium evolves, and it was supposed that electric waves, produced by a special type of oscillator, were propagated through space and thus caused the explosion. But even the ablest chemist could say nothing precise or certain. At last two policemen, who were passing in front of the Hôtel Meyer, found on the pavement, close to a ventilator, an egg made of white metal and provided with a capsule at each end. They picked it up carefully, and, on the orders of their chief, carried it to the municipal laboratory. Scarcely had the experts assembled to examine it, then the egg burst and blew up the amphitheatre and the dome. All the experts perished, and with them Collin, the General of Artillery, and the famous Professor Tigre.

The capitalist society did not allow itself to be daunted by this fresh disaster. The great banks re-opened their doors, declaring that they would meet demands partly in bullion and partly in paper money guaranteed by the State. The Stock Exchange and the Trade Exchange, in spite of the complete cessation of business, decided not to suspend their sittings.

In the mean time the magisterial investigation into the case of those who had been first accused had come to an end. Perhaps the evidence brought against them might have appeared insufficient under other circumstances, but the zeal both of the magistrates and the public made up for this insufficiency. On the eve of the day fixed for the trial the Courts of Justice were blown up and eight hundred people were killed, the greater number of them being judges and lawyers. A furious crowd broke into the prison and lynched the prisoners. The troops sent to restore order were received with showers of stones and revolver shots; several soldiers being dragged from their horses and trampled underfoot. The soldiers fired on the mob and many persons were killed. At last the public authorities succeeded in establishing tranquillity. Next day the Bank was blown up.

From that time onwards unheard-of things took place. The factory workers, who had refused to strike, rushed in crowds into the town and set fire to the houses. Entire regiments, led by their officers, joined the workmen, went with them through the town singing revolutionary hymns, and took barrels of petroleum from the docks with which to feed the fires. Explosions were continual. One morning a monstrous tree of smoke, like the ghost of a huge palm tree half a mile in height, rose above the giant Telegraph Hall which suddenly fell into a complete ruin.

Whilst half the town was in flames, the other half pursued its accustomed life. In the mornings, milk pails could be heard jingling in the dairy carts. In a deserted avenue some old navvy might be seen seated against a wall slowly eating hunks of bread with perhaps a little meat. Almost all the presidents of the trusts remained at their posts. Some of them performed their duty with heroic simplicity. Raphael Box, the son of a martyred multi-millionaire, was blown up as he was presiding at the general meeting of the Sugar Trust. He was given a magnificent funeral and the procession on its way to the cemetery had to climb six times over piles of ruins or cross upon planks over the uprooted roads.

The ordinary helpers of the rich, the clerks, employees, brokers, and agents, preserved an unshaken fidelity. The surviving clerks of the Bank that had been blown up, made their way along the ruined streets through the midst of smoking houses to hand in their bills of exchange, and several were swallowed up in the flames while endeavouring to present their receipts.

Nevertheless, any illusion concerning the state of affairs was impossible. The enemy was master of the town. Instead of silence the noise of explosions was now continuous and produced an insurmountable feeling of horror. The lighting apparatus having been destroyed, the city was plunged in darkness all through the night, and appalling crimes were committed. The populous districts alone, having suffered the least, still preserved measures of protection. They were paraded by patrols of volunteers who shot the robbers, and at every street corner one stumbled over a body lying in a pool of blood, the hands bound behind the back, a handkerchief over the face, and a placard pinned upon the breast.

It became impossible to clear away the ruins or to bury the dead. Soon the stench from the corpses became intolerable. Epidemics raged and caused innumerable deaths, while they also rendered the survivors feeble and listless. Famine carried off almost all who were left. A hundred and one days after the first outrage, whilst six army corps with field artillery and siege artillery were marching, at night, into the poorest quarter of the city, Caroline and Clair, holding each other's hands, were watching from the roof of a lofty house, the only one still left standing, but now surrounded by smoke and flame. Joyous songs ascended from the street, where the crowd was dancing in delirium.

"To-morrow it will be ended," said the man, "and it will be better."

The young woman, her hair loosened and her face shining with the reflection of the flames, gazed with a pious joy at the circle of fire that was growing closer around them.

"It will be better," said she also.

And throwing herself into the destroyer's arms she pressed a passionate kiss upon his lips.

4

The other towns of the federation also suffered from disturbances and outbreaks, and then order was restored. Reforms were introduced into institutions and great changes took place in habits and customs, but the country never recovered the loss of its capital, and never regained its former prosperity. Commerce and industry dwindled away, and civilization abandoned those countries which for so long it had preferred to all others. They became insalubrious and sterile; the territory that had supported so many millions of men became nothing more than a desert. On the hill of Fort St. Michel wild horses cropped the coarse grass.

Days flowed by like water from the fountains, and the centuries passed like drops falling from the ends of stalactites. Hunters came to chase the bears upon the hills that covered the forgotten city; shepherds led their flocks upon them; labourers turned up the soil with their ploughs; gardeners cultivated their lettuces and grafted their pear trees. They were not rich, and they had no arts. The walls of their cabins were covered with old vines and roses. A goat-skin clothed their tanned limbs, while their wives dressed themselves with the wool that they themselves had spun. The goat-herds moulded little figures of men and animals out of clay, or sang songs about the young girl who follows her lover through woods or among the browsing goats while the pine trees whisper together and the water utters its murmuring sound. The master of the house grew angry with the beetles who devoured his figs; he planned snares to protect his fowls from the velvet-tailed fox, and he poured out wine for his neighbours saying:

"Drink! The flies have not spoilt my vintage; the vines were dry before they came."

Then in the course of ages the wealth of the villages and the corn that filled the fields were pillaged by barbarian invaders. The country changed its masters several times. The conquerors built castles upon the hills; cultivation increased; mills, forges, tanneries, and looms were established; roads were opened through the woods and over the marshes;

the river was covered with boats. The hamlets became large villages and joining together formed a town which protected itself by deep trenches and lofty walls. Later, becoming the capital of a great State, it found itself straitened within its now useless ramparts and it converted them into grass-covered walks.

It grew very rich and large beyond measure. The houses were never high enough to satisfy the people; they kept on making them still higher and built them of thirty or forty storeys, with offices, shops, banks, societies one above another; they dug cellars and tunnels ever deeper downwards. Fifteen millions of men laboured in the giant town.

Population, Urbanization, and Congestion

Arnold J. Toynbee

To be catapulted into megalopolis straight out of Arcadia is to be given a shock that may turn an innocent countryman into an urban criminal lunatic. This was borne in on me when I was once hovering, in a helicopter, over the interior of Puerto Rico. From only a few hundred feet up, I was looking down at a choppy sea of jungle-covered hills with, here and there, a tiny clearing, containing just one cottage and one corn-patch. If I had been born and brought up in one of these secluded tropical homesteads, out of sight and hearing of the rush and roar of the modern urban world, how should I have felt if, under the spur of economic pressure, I had suddenly been transported to the East Side of New York City? Might I not have lost my moral bearings when, like a palm tree caught in a hurricane, I had been torn up from my social roots? I now understood why, in New York, some of the Puerto Rican immigrants make awkward neighbours for the better-acclimatized older inhabitants. I also appreciated the wisdom of the Puerto Rican Government's policy of seeking to create industrial employment at home for the redundant rural population of the island by offering to United States corporations attractive financial inducements to set up branch-factories in Puerto Rico. It might not be easy for a Puerto Rican peasant to accustom himself to living and working in San Juan; but the transition from agricultural to industrial work would at least be less upsetting for him if he were given the opportunity of making it without having to leave his native shores.

From *Change and Habit*, by Arnold J. Toynbee (Oxford, 1966), pp. 204-211.

But how is the Puerto Rican peasant, and how are we, to make urban life tolerable for ourselves when San Juan, as well as Princeton, N.J., and Los Angeles and Athens, has been engulfed in the world-wide city of the future? Our most urgent task is to rescue mankind from the shanty-towns, like those now encompassing Arequipa and Baghdad, in which more and more millions of human beings are being dumped like worn-out cars on a mammoth-size rubbish heap. Happily, this sordid menace to life, liberty, and the pursuit of happiness has already evoked some creative planning and building which gives hope for the future.

On my first visit to Calcutta and Karachi after the partition of the former British Indian Empire, I was horrified at the misery in which the millions of 'displaced persons' were living (if such existence can be called life) in the slums of these two cities. On my next visit to Karachi, not many years later, I found that the shanty-towns among the mangrove swamps had been cleared away and that their former occupants had been rehoused. The site of this rehousing had been, in its virgin state, a bare dry desert which, in its own way, had been as forbidding to look at as the swamps, though it was, of course, not pestilential but salubrious. I now found this former desert clothed, as far as the eye could see, in a cluster of new satellite cities. The brick-work was simple (Pakistan has no money to spare for luxury building), but the result that had been achieved at a minimum cost in terms of money was satisfying in terms of human needs. Each of the new cities had been articulated into a number of distinct hamlets, and each of these hamlets had been kept within a human size—that is to say, within a size that would allow the inhabitants to be personally acquainted with their neighbours. Each hamlet had its own store, school, mosque, and washhouse, and all these public facilities were within walking distance of the people's new homes. As a result, the displaced persons' lives had been re-humanized. They had not, it is true, been repatriated to their original homes, but they had been rescued from the shanty-town and had been rehoused in circumstances in which they were evidently striking new roots.

Who, I asked, had planned these inspiriting new cities for the Pakistani Government? The designer, I was told, was Mr. C. A. Doxiadis of Athens, Greece. I was eager to meet the man who had done so human a service for so many human beings in distress; and, when I did meet him in Athens, later on, my first question to him was: 'Mr. Doxiadis, how did you discover what the living-conditions are that give displaced persons a chance of making a new start in life?'—'I did not have to discover this,' Mr. Doxiadis replied; 'I knew it by experience, because I grew up with the problem. I and my family are displaced persons ourselves. My father was a doctor in the former Greek community in the town of Stanimaka in Bulgaria.' Mr. C. A. Doxiadis and his brothers had, in fact, to make a new start in life in Greece. Mr. C. A. Doxiadis has responded to this challenge by inventing a new profession and by using the experience and resources that this has given him for founding a new science—or rather, a new art. His profession is truly a new one; for 'town-planning,' in the former rather perfunctory meaning of the term, would be an inadequate name for it. While Mr. Doxiadis has planned for Karachi and for Philadelphia, Pennsylvania, he is also planning for megalopolis; and, with the population explosion and the depopulation of the countryside always in mind, he is planning, not just for the next decade, but for the next century. He is working out a configuration for megalopolis that will allow for continuous expansion without producing an intolerable congestion round a static centre. Mr. Doxiadis has named this new art of his 'ekistics.' This imaginative approach and long-term view open up for future generations a prospect of mastering megalopolis instead of being victimized by it.

Yet, however vigorously and skilfully the new art of ekistics may be developed by Mr. Doxiadis and his disciples, congestion, to a hitherto unknown degree, is bound to be the consequence of the imminent tripling or quadrupling of the size of our planet's population. We are fast moving into a so far barely imaginable new world in which the largest surviving open spaces will be the airports for supersonic aircraft. There will be no room left for traffic on the surface. Goods that are too heavy to be carried by air will have to be transported underground. There will be no room left, either, for agriculture on land; streets and houses will occupy every acre of terra firma. Food production, if not fibre production, will be driven out to sea. We shall exterminate the sharks, cat-fish, and other predatory marine wild life, and shall stock the sea with flocks of herrings and with herds of whales, as we are already stocking our lakes and rivers with edible fresh-water fish. Our iron ration will be domesticated plankton.

While travel on business will be on a greater scale than ever before, travel for pleasure will probably tail off. The incentive for sight-seeing will disappear in a standardized world; for what will a man gain by visiting the Antipodes when there will be nothing to be found there but the ubiquitous megalopolis that will have been his starting-point and his accompaniment *en route*? If restlessness still drives him to undertake a journey that can no longer minister to his curiosity, the public authorities will intervene. For, though aeroplanes will abound, what will they be among so many? The authorities will be chary of issuing travel-permits. 'Is your journey really necessary?' they will ask, as they asked in Britain during the Second World War. 'Need you travel in person?' they will add. 'Cannot you travel vicariously, instead, by using the telephone, telegraph, radio, or TV?' Except for strictly necessary business purposes—and these will whirl people round the globe in a trice—the inhabitant of megalopolis will be more and more rigorously confined within the bounds of his own human-size quarters of the regional ward. What kind of a life will he and his neighbours be able to make for themselves within these oppressively narrow physical limits?

Cities in Trouble: Frederick Law Olmsted

Stewart L. Udall

Proud, cruel, everchanging and ephemeral city
To whom we came once when our hearts were high,
Our blood passionate and hot,
Our brain a particle of fire:
Infinite and mutable city, mercurial city,
Strange citadel of million-visaged time—
O endless river and eternal rock,
In which the forms of life
Came, passed, and changed intolerably before us!
And to which we came, as every youth has come,
With such enormous madness,
And with so mad a hope—
For what? . . .

Thomas Wolfe
From "The Ghosts of Time"

The urbanization of America has been a striking trend of the twentieth century. In Theodore Roosevelt's time we were still a predominately rural people; now we are predominately urban and we are become more so by the day.

Our cities have grown too fast to grow well, and today they are a focal point of the quiet crisis in conservation. The positive appeals of the modern city—the stimulating pageant of diversity, the opportunities for

From *The Quiet Crisis*, by Stewart L. Udall (New York, 1963), pp. 159-172.

intellectual growth, the new freedom for individuality—have been increasingly offset by the overwhelming social and economic and engineering problems that have been the by-product of poorly planned growth.

Under explosive pressures of expansion there has been an unprecedented assault on urban environments. In a great surge toward "progress," our congestion increasingly has befouled water and air and growth has created new problems on every hand. Schools, housing, and roads are inadequate and ill-planned; noise and confusion have mounted with the rising tempo of technology; and as our cities have sprawled outward, new forms of abundance and new forms of blight have oftentimes marched hand in hand. Once-inviting countryside has been obliterated in a frenzy of development that has too often ignored essential human needs in its concentration on short-term profits. To the extent that some of our cities are wastelands which ignore and neglect the human requirements that permit the best in man to prosper, we betray the conservation ethic which measures the progress of any generation in terms of the heritage it bequeaths its successors. The citizens of our cities must demand conservation solutions based on the principle that space, beauty, order and privacy must be integral to its planning for living. As long as those designers and planners who might help us create life-giving surroundings remain strangers at the gates we will not create cities which are desirable places to live. Today, "progress" too often outruns planning, and the bulldozer's work is done before the preservationist and the planner arrive on the scene.

Between 1950 and 1959, while our cities' populations increased by only 1.5 per cent, the population of our suburbs increased by 44 per cent. Even this flight to the suburbs—in part a protest against the erosion of the urban milieu—has had its element of irony, for the exodus has intensified our reliance on the automobile and the freeway as indispensable elements of modern life. More often than not, the suburbanite's quest for open space and serenity has been defeated by the processes of pell-mell growth.

Many mental-health experts have offered evidence of the corrosive effects on the human psyche of the unrelieved tension, overcrowding, and confusion that characterize city life. There is a real danger that the struggle with ugliness and disorder in the city will become so all-consuming that man's highest and most specifically human attributes will be frustrated.

The prime business of those who would conserve city values is to affirm that such human erosion is unnecessary and wasteful; that cities can be made livable; that with proper planning the elements of beauty and serenity can be preserved.

Urban America has had, in Frederick Law Olmsted, its own conservation prophet and master planner. Lewis Mumford once called Olmsted "one of the vital artists of the 19th Century"; he has had no peer in the United States as a community designer.

Olmsted did his pioneering work in a period when the need for public playgrounds was not recognized and the art of city planning was largely ignored. In 1859, while Thoreau was noting in his journal that each town should have a miniature wilderness park ("a primitive forest of five hundred or a thousand acres where a stick should never be cut for fuel, a common possession forever"), Olmsted was already developing a design for a proposed "central Park" in the heart of Manhattan Island, and thus began his career as a conserver of higher values in the city.

The key to Olmsted's genius was his insistence that all land planning had to look at least two generations into the future. His versatile talents found many outlets. For a time during the Civil War he headed up the United States Sanitary Commission, the forerunner of the American Red Cross. As a temporary resident of California, before John Muir arrived, Olmsted played a leading role in the enactment of the Yosemite park bill signed by President Lincoln in 1864—and some of his park-management ideas anticipated the subsequent standards of the national park system. He did land-use planning for San Francisco, Buffalo, Detroit, Chicago, Montreal, and Boston. He was commissioned to landscape the grounds for the Capitol and the White House in Washington, and he designed the 1893 World's Columbian Exposition held in Chicago.

By orderly planning and provision for abundant natural areas, Olmsted believed cities could keep sufficient breathing and playing space to allow continual self-renewal. He proposed that part of the countryside be preserved within each city. His Greensward plan for Central Park was, as he later wrote, designed to "supply to the hundreds of thousands of tired workers, who have no opportunity to spend their summers in the country, a specimen of God's handiwork that shall be to them, inexpensively, what a month or two in the White Mountains or the Adirondacks is, at great cost, to those in easier circumstances. The time will come when New York will be built up, when all the grading and filling will be done. . . . There will be no suggestion left of its present varied surface, with the single exception of the few acres contained in the Park."

Central Park was a dreary stretch of rock and mud when Olmsted took charge. Working with nature, he tried to visualize and anticipate the growth patterns of a great metropolis. All effects in the park—trees, mounds, ponds, paths, meadows, groves—were carefully composed with an eye to creating life-promoting surroundings. Shrubbery screened out the works of man, and Central Park became an oasis where urban man could refresh his mind and soul.

Olmsted doggedly fought off the politicians and the well-meaning promoters who wanted to install on the grounds a stadium, a theater, a full-rigged ship, a street railway, a race track, a church, a permanent circus, a cathedral, and a tomb for Ulysses S. Grant. In each victory he affirmed the primacy of park purposes and strengthened the idea that some parkland had to remain inviolate.

As his vision broadened, Olmsted became more than a mere planner of parks. He saw that urban design should include the whole city and provide diverse and continuous enclaves of open space, green gardens, and public playgrounds. Had he been able to win support for his bold conceptions, the shape of many of our cities might be different today. His aim was to suit the city to the individual, and not vice versa, and perhaps his achievement of a healthy balance between the works of man and the works of nature in an urban setting is his most durable monument.

Despite the success of Central Park, the city fathers did not adopt Olmsted's most farsighted recommendations. Open space and elbow room cost money even then, and in a period of hectic growth the vision of a Frederick Law Olmsted was too advanced for the apostles of "progress."

* * * *

The Olmsted ideas are still applicable, and even today most American cities have unrecognized opportunities, both within their corporate limits and on their fringes, to save large and small Central Parks for the future. Every well-conceived urban redevelopment project offers an opportunity to create green spaces in the central city and avenues of action are open to conservationist city leaders and citizen groups who set store by civic beauty and are willing to levy sufficient taxes for environment preservation. They should invite the Olmsteds of our time to participate in the redesigning of our cities.

No attitude is more fatal today than the belief of some local leaders that economic salvation lies solely in getting new property on the tax rolls. Central Park cost something over $5,000,000 in the 1850's. It is worth billions today, and much of its value lies in its ad valorem and esthetic enhancement of surrounding property. Money spent on a properly planned environment is an investment not only in future taxability, but in the physical and mental health of the residents—their efficiency, their general well-being, and their enjoyment of life.

If we are to create life-enhancing surroundings in both cities and suburbs, the first requirement is the power to plan and to implement programs which encompass the total problems of metropolitan regions. Air and water pollution, recreation, and provision for adequate mass transit are region-wide problems, but in most areas action is hampered by legal impediments which actually prevent regional planning. As long as each city, county, township, and district can obstruct or curtail, planning for the future cannot be effective. The cities and metropolitan areas that are devising new political institutions for regional planning are today's pioneers of urban conservation.

But local governments still hold the key to planning. Many zoning boards are as important as the courts. Zoning regulation should not merely prevent the worst from happening . . . it should encourage positive action to provide esthetic opportunity for the present and future while preserving the history of the past.

* * * *

Both zoning and tax incentives may be used to stimulate another method of saving open space: cluster development. By clustering homes and designing smaller yards immediately surrounding them, a developer can provide the customary number of dwellings per acre and yet preserve as much as one-third of a given area in its natural condition. This offers residents pleasant networks of wooded walks, stream-side parks and other recreation areas. This concept in a sense marks a return to the village greens and town commons of America's colonial period. Some communities in the United States and Canada have encouraged this kind of preservation of natural areas by requiring that a proportion of the total area in any new subdivision be dedicated to open space.

Another innovation in land planning is the conservation easement. In the seacoast county of Monterey in California aroused citizens secured the enactment of a state "open space" law in 1959, enabling cities and counties to purchase property, or easements on property, for the purpose of preserving pastoral landscapes. In purchasing an easement on open land, the public agency acquires a "right" from the owner, but otherwise leaves him full ownership and the property remains on the tax rolls. The "right" might be simply that the land remain in its natural state, as in the case of scenic easements. Usually an owner sells the right, but sometimes he may donate it voluntarily. In scenic Monterey County, California, many landowners have given voluntary easements covering thousands of acres, including parts of the spectacular shore line at Big Sur. In return for this landscape preservation, donors are protected against rising assessments that would force them to subdivide or sell.

Measures similar to the California open-space law have been adopted by several other states, including New York, Massachusetts,

Maryland, Connecticut, and Wisconsin. The last-named state, for example, has purchased scenic easements at very low cost to preserve countryside along the Great River Road down the Mississippi. In New York State a different kind of easement provides access rights for fishermen along many miles of privately owned trout streams.

There are other useful tools for land planning. Agricultural land, for example, can be purchased by a public agency and leased back to the former owner with the proviso that its pastoral character be maintained. Ottawa, the Canadian capital, has pioneered in the use of this technique. In order to control its growth pattern the National Capital Commission of Canada purchased a semicircular belt of farmland and open space to the south of Ottawa. The inner margin of this green belt is about six miles from the center of the city and the belt itself is two and one-half miles deep and embraces some 37,000 acres. Most of the original farming and open-space uses continue as before.

The city is bounded on the north by the Ottawa River, beyond which a wedge-shaped extension of Gatineau Park eventually will comprise 75,000 acres. Although some phases of Ottawa's park and open-space program date from the turn of the century, the greater part of the master plan was developed by the French designer, Jacques Greber, after World War II. Ottawa today is a metropolitan area that would be regarded as a model by Olmsted himself. About 300,000 people live next door to more than 100,000 acres of superb parkland and green space that provide a permanent corridor of natural beauty for the capital city of Canada.

As the result of a more recent master plan, the county which surrounds Phoenix, Arizona, has laid out, as a land bank for the future, a peripheral group of regional parks embracing some 75,000 acres.

Pioneer open-space legislation has been enacted in Connecticut, which encourages communities to lower taxes on open land, to buy land or purchase interest in it, and to lease back purchased land subject to restrictions—all backed up by financial help and by the power of condemnation. As in Massachusetts, community conservation commissions have been established and given broad powers to protect and enhance their environments.

Vital as parks and open spaces are, they alone will not save our urban areas. Inevitably, cities will continue to be predominately manmade, and urban conservation must include the artificial as well as the natural. The most beautiful American cities are notable not only for their natural landscapes but also for the design and organization of the buildings and projects that make up the total environment. San Francisco, known for its superb natural setting, has enhanced its hills and bay with some of the most inspiring bridges constructed since the Romans demonstrated that bridge-building was an art as well as a science. Wisconsin's noted native son, Frank Lloyd Wright, argued throughout a long lifetime that buildings should blend with landscapes, and not vice versa; and Madison, his state's capital city, located among beautiful lakes and moraine hills, has made the most of its unusual terrain. Were he alive, Thoreau would have equal admiration for the 1,200-acre arboretum on Madison's outskirts, and a recently acquired 700-acre marsh within its city limits.

Such cities as Santa Fe, New Orleans, and Boston have been set apart by the distinctive way in which their historical sections have been preserved; and the renewal of downtown Philadelphia is an exciting example of both urban rebuilding and the conservation of historical landmarks.

Thanks to L'Enfant's grand design of 1791, Washington is one of the few large American cities planned from its inception. It also is one of the few large cities that has avoided skyscraper blight; in order that the Capitol, the Washington Monument, and other federal buildings should not be overwhelmed by taller structures, the height of buildings has been rigidly controlled. As a result, Washington has retained a green and spacious appearance, and its physical scale puts a premium on human values. The city's skyline and its vistas are dominated by the domes, monuments, and cathedrals that declare the aspirations of the American people.

City planning should put people first. Autos, freeways, airports, and buildings should not be allowed to dominate a city; each must take its own place in a balanced environment along with trees and parks, playgrounds and fountains. Just as there are certain areas from which skyscrapers should be excluded, so there should be more places where the automobile is off limits. Well-placed malls, plazas, promenades, and gardens can become oases inviting delight and giving a sense of order to living.

The crowding of our urban regions has caused us to look with new interest on such "useless" natural areas as marshes and swamps. A few years ago the suggestion that swamps might help save our cities would not have been taken seriously; but residents of Washington, D.C., Philadelphia, Madison, and some New Jersey communities have established large areas of swamp and marshland as permanent nature sanctuaries. It takes a perceptive eye to see the miracles of life in the woodlands and bogs where our forefathers would have seen only another opportunity to subjugate nature.

In all phases of city development we need to give free rein to imaginative designers like Eero Saarinen, whose Dulles International Airport near Washington is attracting worldwide acclaim; or Nathaniel Owings, the main architect of San Francisco's glass-clad Crown-Zellerbach Building, located in its own park in downtown San Francisco. Public buildings, which are too often the scenes of shallow triumphs of penny-pinching officials, should set the pace in architectural design, in landscaping, and in the use of painting and sculpture.

From the standpoint of urban design, the size of many of our large cities has already reached the point of diminishing livability. Just as there is an optimum density of population within a given area, so there must be an optimum physical size for cities.

Long before universal double-decking and the overuse of vertical space make congestion intolerable, we must give more attention to the only practical alternative: the creation of new cities. The best of our "industrial parks" may point the way. Tax allowances and other incentives now encourage industries to locate in new areas in accordance with a master plan for land use.

Borrowing from the Ottawa pattern, some planners have visualized large urban constellations involving industrial parks, clustered housing, plentiful recreation areas and extensive green belts. These new cities should have a community life of their own and become creative centers of commerce and culture, "park cities" that would give priority to community living.

Innovations in technology are sure to provide opportunities for new kinds of urban planning: the development of nuclear reactors as a safe, cheap source of power; advances in air transportation; and the perfection of high-voltage, long-distance power transmission lines which will enable us to transmit electrical energy economically anywhere in the country will all aid the planners of tomorrow's cities. Together, these techniques will reverse the age-old process of locating cities only near waterways or along main arteries of commerce. Many planners are convinced that the principal hope for accommodating a much larger population in this country without impossible crowding lies in the development of new cities which will range in population from 30,000 to 300,000 people.

* * * *

There is an unmistakable note of urgency in the quiet crisis of American cities. We must act decisively—and soon—if we are to assert the people's right to clean air and water, to open space, to well-designed urban areas, to mental and physical health. In every part of the nation we need men and women who will fight *for* man-made masterpieces and against senseless squalor and urban decay.

Good Buildings and the Good Life

An Interview with Allan Temko

Q: Mr. Temko, what relationship exists between the architect and the city in this country?

TEMKO: At present there is almost none. The architect cannot do more than an individual building or group of buildings in a city. He might get an urban redevelopment project, perhaps even a large one. But none has ever been assigned a Brasilia, like Oscar Niemeyer, or a Chandigarh, like Le Corbusier. An example of the lack of coordination between the architects and the city and between the architects themselves is shown in what has been happening to Rockefeller Center. Rockefeller Center was the work of a group of architects in the nineteen-thirties. It was a complex of buildings in which open space had been planned as an integral part of the composition. It was a good start but somewhat overrated because only seventeen per cent of the site had been left open —not much in view of the congestion created by the Center itself. But it was a start and it was something new in the history of private development.

Now, however, we have regressed. The Time & Life Building, though not visually an extension of Rockefeller Center, is considered a part of it, and it is only one of four huge skyscrapers that will be on the four corners adjacent to the Center. The Columbia Broadcasting System and Equitable Life Assurance occupy two of the other corners. The fourth was bought from William Zeckendorf by Uris Brothers—they put up much of the commercial hackwork in New York. The architects involved are above average. Eero Saarinen designed the CBS building.

From *The Center Magazine*, II:56-63 (November 1969).

Harrison & Abramovitz did Time & Life. Skidmore, Owings & Merrill did the Equitable Building. The smallest of these skyscrapers will be the CBS building at, I believe, thirty-eight stories. But these four corporations—though two of them have high cultural pretensions—could not create one plaza between them. They were not able to put Sixth Avenue (the Avenue of the Americas) underground for a couple of blocks. They could not re-route traffic. They could not create a city-scape. All they could do was add to what Lewis Mumford calls "solidified chaos": Manhattan at its worst, skyscrapers going up without any relationship to human needs and aesthetic sensibilities.

Q: In this "could not" is there also an element of "would not"?

TEMKO: They could have, but there was no political means to compel them to do it.

Q: You mean, there was no urban authority to say, "You must do this, or that"?

TEMKO: Yes. There was no municipal authority or even any municipal guidance, except standard code and zoning based on speculative land values.

Q: Is this typical of how our large American cities have grown through the years?

TEMKO: Yes, it helps explain the chaos. Our federally sponsored redevelopment projects are now organized, but I find most of them disappointing.

Q: Why? Is it because the people who have ultimate control over the projects lack the experience and education to make proper decisions? That is, our borough presidents, city councils, and aldermanic bodies have the authority and responsibility for redevelopment, but no special or expert competence to exercise that authority to good effect?

TEMKO: A city like New York is completely out of control, both at its center and at its periphery. Even if it had a much finer municipal government, it would take a full generation to arrest the present disintegrating tendencies as far as urban life is concerned.

Q: What is wrong? Lewis Mumford, for example, says that our cities lack the humanist dimension and orientation, that in many instances they frustrate rather than nourish human relationships.

TEMKO: The modern metropolis in many respects is inhuman. There is over-congestion, an over-concentration of people. The automobile is out of control. In New York it is next to impossible to drive; this costs billions of dollars in wasted man-hours. We have no way of organizing residential and industrial patterns. Simple zoning is not the answer; it is piecemeal. I.B.M. moved its headquarters to Westchester County. They got some wonderful architecture. Saarinen, for example, did their research building. But this big corporation moved out of Manhattan and into Westchester County completely independently, without any coordination. Until such moves are coordinated and people decide what things should be in the city and what should be outside the city, we are going to be in trouble.

Q: Are you saying that we need not only greater artistic competence in the development of our cities but also a greater measure of authority?

TEMKO: Yes. The very word "civic" has implications which were clear in ancient societies. The Athenian citizen, for example, took large responsibility for the physical appearance of his city.

Q: I suppose there would be an immediate outcry if some municipal or metropolitan authority were to say to, for example, I.B.M.: "Wait, you can't move until we see whether such a move fits in with our over-all urban program." In some quarters at least, I can imagine such restraint would be labeled "authoritarianism."

TEMKO: The example of I.B.M. is one that may, in fact, work out very well. The point is, it may *not* work out very well, and there was no prior coordination of such a move in terms of what is good not only for I.B.M. but also for New York City and Westchester County. Freedom does imply responsibility. But even the most responsible corporation cannot now coordinate its moves with either the city or society as a whole because the means for such coordination do not exist. There is a plethora of laws, but they are working out badly.

Santa Barbara has very strict regulations on architecture, intended to make the city a Spanish-colonial stage set. In some ways it has worked out, but it has not become a visual utopia. It has, for example, the same hideous gas stations you will find in Indianapolis. Look around the San Francisco region. It is one of the most beautiful natural environments in the world and what have we done to it? We have polluted the Bay. We have ruined the hills. We have ruined the atmosphere. We have no metropolitan government. We don't even have a Bay region authority or plan. Dade County, Florida, has a metropolitan government, and Chicago has a metropolitan regional plan, but a plan with no teeth in it. Washington, D.C., has a superficially impressive plan, but none of the controls necessary for its success.

Q: Is there some middle way between iron control on the one hand and no control at all on the other?

TEMKO: The Russians have an authoritarian system of planning which does not seem to be working out much differently from ours as far as overgrowth is concerned. But there is a third approach found in the Scandinavian countries and in Great Britain. The postwar British Labor government started a very bold environmental program which has been continued by the Conservatives. It includes the green belt around London, the New Towns, and very carefully designed housing. These British developments have faults, but they are so far ahead of what we have done that no real comparison is possible.

Q: What are some of the characteristics of the ideal city from the standpoint of human relations?

TEMKO: We must remember that the city is made up of all kinds of individuals who wish to live in different ways. A city should provide that variety. Lake Meadows in Chicago is a significant renewal project. It was built in the nineteen-fifties by Skidmore, Owings & Merrill and was financed by the New York Life Insurance Company. It is the largest exemplification of Le Corbusier's planning ideas that I know in this country. The Lake Meadows site was one of the worst slums in the world. Certainly it was the worst I have ever seen—and I have been in China. They decided to raze it. Here was a case where they worked on a large enough scale but, architecturally and artistically, they started rather timidly. They put up five rather nondescript towers about twelve stories high in an X formation—two towers on either side and one in the center. They were stubby in outline and rather grim. They suffered from "projectitis." Ninety-nine per cent of the first occupants of these five buildings were Negroes. The other one per cent were young architects who moved in because of an intellectual commitment. A second group of buildings went up and this was much bolder. Here were four huge slabs, each about twenty-two stories high. They are not great, but they are excellent buildings. Two of them are about four hundred feet apart; the other two stand somewhat out and behind them, farther from the lake, and about six hundred feet apart. So you have a magnificent funnel of fresh air coming off the lake, affording great relief in the summer. These four great slab apartment houses are set in a green park: Le Corbusier's idea of skyscrapers-in-parks. In winter, of course, it is rather forbidding, with the Chicago weather. The architects originally wanted glazed, heated passageways for the winter, but they did not get them because of the cost.

Q: What kind of tenants did they get for this second group of buildings?

TEMKO: I find this whole project interesting for its sociological implications. It quickly became apparent that this second group was the best housing for the money in Chicago. Whites started to move in. They included not only architects, who from the beginning saw these apartments as excellent but also army sergeants, television repairmen, civil servants. The proportion of whites to Negroes grew to 20-80, then 30-70, and in a new series of buildings put up nearby by less gifted architects—a project called Prairie Shores—the ratio was about 50-50. The question then arose: Suppose these housing projects became totally white? I asked Fred Kramer, the Chicago financier who is actively involved in these developments, whether he had contemplated doing anything to maintain a percentage of Negroes. He said he hoped that this would take care of itself normally, that it would stabilize itself according to the proportion of Negroes and whites in the rest of the city. Recently they have added a luxurious apartment house to Lake Meadows' great group of four buildings and the rents are as high as any in Chicago, even on the North Shore. The tenants include both Negroes and whites, some with very high incomes.

Q: Is it simply provision of green space between the buildings that makes them desirable?

TEMKO: The people find it attractive to live there. If you have young children, however, you probably do not want to live nineteen stories in the air. At Hyde Park, Zeckendorf's project near the University of Chicago, there is a drastically different approach to urban renewal, in which you have two big apartment houses by an excellent architect, I. M. Pei, but very close together, rather needlessly formal, and surrounded by an elliptical roadway, a "gasoline alley," so that traffic has not been removed. And in Hyde Park if you are on the inner court between the two great towers you can see the people across the way in their underwear unless they draw their blinds. One advantage of towers set very far apart, as at Lake Meadows, is the privacy afforded. Hyde Park also has little squares with town houses which are supposed to provide intimacy, like the little squares in Bloomsbury in London. But in Hyde Park they are a failure because the automobile is everywhere, taking up the center of these little courtyards and squares, which should be green. To me the present housing is inferior to what previously existed when the neighborhood was made up of free-standing one-family houses.

Q: What about transportation? Isn't that one of the central unsolved problems of urban living?

TEMKO: The auto problem is out of control at present, again because of a failure of politics. I consider this primarily political, though most people do not. I think we can act only through government on all levels, federal, state, and local, to get the problem of transportation under control. We are subsidizing the automobile industry. We have a fantastic national investment in automobiles. We change cars every year. We are building highways that are a feast for the bankers as well as the contractors. Meanwhile, railroads die and you cannot get money or support for alternative rapid-transit systems in our big metropolitan areas. The San Francisco Bay region has an excellent rapid-transit plan. Originally it was intended for nine counties; as finally presented it included five counties. Of those five, two counties have already withdrawn. Marin County withdrew because of a technical difficulty: a panel of engineers found that the Golden Gate Bridge could not carry trains. But San Mateo County withdrew because of irresponsible politicians, who acted against the will of many citizens, perhaps a majority.

Q: Ideally, there should be a diversity of transportation methods in our cities, should there not? And yet there is not much that can be done if people insist on driving their cars between the suburbs where they live and the central city where they work or go to school.

TEMKO: Part of this is educational. The Americans get a lot of psychological satisfaction, or maybe it is only an illusion of satisfaction, from their cars. The auto is a real convenience, too. The American loves to go from one place to another without changing his transportation. I don't know what the answer is, but it is a problem that must be attacked on several levels. And in many cities we do not have any good alternative to the available system of transportation.

Q: It is said that the freeways and expressways in some cases have aggravated rather than relieved traffic congestion. At certain hours in New York or Los Angeles, for example, traffic on these roads slows down to five or six miles per hour.

TEMKO: Yes, and that is slower than the rate of the old horse cars. I think it was much faster to take a trolley car in Chicago thirty years ago than it is to drive your car there now. When people see the implications of some of these things they rebel.

Q: Isn't it one of the complaints of the architects that they are not called in soon enough on the large urban renewal projects in our cities?

TEMKO: Yes. The Golden Gateway project in San Francisco was laid out by an architectural firm, but other things conditioned what it could do. For example, a freeway runs very close to the site where the best type of modern housing is supposed to be built. Another bad freeway borders the renewal area in southwest Washington, D.C. Why those site problems could not have been solved architecturally at an early stage, no one knows. The engineering mentality always resists this sort of intrusion by the artists and architects.

Q: Have there been any "success stories" in the relationship of architects to urban planners in any of our cities? Have there been any instances in which competent architects have been called in at the beginning of long-range city planning, have actually planned and carried through a large-scale project?

TEMKO: Of course in the nineteenth century artist-designers were called in to work on a grand scale. Frederick Olmsted, the landscape architect, was one. I recently read the memoirs of T. Jefferson Coolidge, a Boston Brahmin, who spent many days on horseback with Olmsted, laying out the magnificent parks around Boston. I wonder how many of our civic leaders today would take the time, and, furthermore, would have the same keen sense of responsibility for their cities.

Q: We have talked about a number of elements that help to make a good city, such as the provision of green space and the control of the transportation problem. You also mentioned parks. Is there a general rule that the more space a city sets aside for parks the better city it will be?

TEMKO: The question is, what makes a good city? Today there is much talk, especially by people who have not been to Europe, that we do not need the great green spaces, that these destroy some of the urban qualities, particularly intimacy. What such people want is the nineteenth-century American city "cleaned up" a little. What they do not understand are the principles of European urbanism. When people over-value Union Square in San Francisco, for example, it is because they have not really studied the great European squares and analyzed their superiority. In those squares traffic may be banished altogether. Think of St. Peter's Square in Rome—Union Square doesn't seem much when compared to it. And even though the Place de la Concorde is cut up with traffic, it is very superior to Union Square. Then there is the Louvre, with the gardens of the Tuileries, which, in modern terms, can be seen as virtually a super-block of fifty acres in the heart of Paris, and with a river next to it. The great parks of central London probably total several hundred acres. You could throw the financial district of San Francisco into them and lose it. Golden Gateway, our huge redevelopment project on the edge of the financial district here, is only forty-odd acres.

Q: Is it true that one of the things that make European city streets interesting and "human" is the existence of outdoor cafés as well as the parks and squares?

TEMKO: Yes, and these matters go deep. In this country we are all "at home" in the suburbs. The American counterpart of the Parisian who lives on the Left Bank now often lives in the suburbs, in a very fine house. We are very comfortable in the suburbs, and therefore why should we go into San Francisco or Chicago or New York? In Paris, you are likely to have something nicer outside than in your flat, especially if you don't have much money. Take the Luxembourg Gardens, a tremendous thing to have so close to the heart of a great city. And actually to see them is to know what a great city can be. In this country I don't think we have nearly enough parks. In this I would not discount Le Corbusier's skyscraper-park concept until it has been tried.

Q: It is difficult, I suppose, for people to demand something better than what they have had if they have never been exposed to that "better," so someone has to take the lead in initiating these urban reforms.

TEMKO: This demands, among other things, an educational approach. An architect in Berkeley, Lois Langhorst, is trying to develop a course in the Berkeley school system, thus far for girls only, as a substitute for conventional home economics, in which they will learn something about good design rather than just how to bake a cake or sew a dress. Sir Herbert Read insists that it is all a matter of public education.

Q: I suppose it is true that the architecture of an age necessarily reflects the age. If our architecture is disorganized and inchoate, perhaps it is because our age is that way too. On the other hand, medieval architecture . . . reflected an order and a discipline and the spiritual confidence that were in the air at the time.

TEMKO: I believe that as far as our environment is concerned we have anarchy. Henry Adams characterized the two periods as "medieval unity" and "modern multiplicity." Yet the modern age has certain unities, one of them being industrialism.

Q: If we could develop a personalism as strong as our industrialism, some of our urban and architectural problems might be solved.

TEMKO: The primacy of the person, yes. If people can be educated and have decent places in which to live, and if we can prevent war, the greatest of all the destroyers of our cities, and if we can marshal all our resources in an orderly, intelligent way, think of the potential benefits to people as individuals.

Architecture of course is only part of the problem of the cities. Conceivably we could have a great city of mediocre buildings. It might be a happy place in which to live. And you might have a beautiful city that is not a happy city. Florence has not always been a happy city.

Q: Yet I suspect it would be difficult for anyone to be much exhilarated by mean buildings.

TEMKO: The architectural quality of individual buildings is perhaps less important than the over-all quality of the urban ensemble. For example, if you look at the Parisian boulevards, what you have is not building façades but street façades. You have those neutral gray walls stretching in baroque perspectives of orderly street fronts. At the end of these neutral perspectives you see a monument that may be good or bad architecturally. It might be the Opéra, which is rather a masterpiece of ostentation. Or you might see the Odéon, or the Arc de Triomphe. So the city suddenly comes into its own by the self-effacement of most of its buildings. The height of the buildings is still rigidly controlled in Paris. And up to now they have shared the same material —stone; it is a masonry city. In Paris, too, many functions will be found in one neighborhood—residential and commercial functions—which is good. The Palais-Royal, which was once a palace, is made up of a number of marvelous shops, restaurants, and a theatre on its lower level, which in places open in arcades; then, above, there are apartments. Colette lived there, Jean Cocteau lives there. It is a wonderful place to live. The Palais-Royal is an architectural entity. There is no auto traffic in its great court, which is like a park in a city. It is a super-block, but a well-designed one, and that was three hundred years ago.

Q: I don't like to labor the traffic problem, but it seems to me that what to do with our cars remains one of the great unsolved problems in the general problem of developing beautiful and vital cities.

TEMKO: The car is an economic waster, a space eater, and a psyche ruiner. More than half the prime space in what has been downtown Los Angeles is taken up now by freeways, parking space, and garages. In San Francisco, twenty-five per cent of downtown is used for parking, and if you add the acreage covered by streets and the freeways on the periphery, perhaps half the downtown area is given over to the automobile. In other cities, certain streets are closed to traffic and it is very pleasant to be able to step off a curb and not have to dodge cars. How nice it is in London where you can walk from Piccadilly down to the great system of parks that stretch from Westminster Abbey to Buckingham Palace and beyond. There you see British businessmen stretched out on the grass taking a rather tense nap. But they use their parks. This is a mark of a great civilization.

I don't agree with Jane Jacobs' idea that we need traffic-filled streets. She loves the streets of Greenwich Village. She probably has not seen the streets in Arles or Aix-en-Provence, which are too narrow for cars and so there is no real auto traffic. Why do people love to go to Venice today? Because Venice has no cars. Why was Bermuda such a wonderful, restful resort before the cars came there? If we got rid of all cars, if someone put them all in a huge crate and dumped them into the ocean, the relief would be magnificent. Of course, we would be left with our sprawling civilization and the problems of communication. How would we do it? Well, I think we would have to rebuild our cities.

Q: What hopes do you have that the city in America, which has so great an effect on human values, hopes, and potentialities, will rebuild itself along these personal axes, rather than along the impersonal, technological, and disorganized axes of the past? Granting that such rebuilding may take several generations, are you hopeful that we will head right?

TEMKO: First of all, let me say that I see no incompatibility whatever between humanism and the rational use of technology. What we find mechanical and disorganized in cities today is due to the misuse of technology. Quite simply, man has failed to use machines with sufficient boldness, sensitivity, kindness, and conviction. Yet as Mies van der Rohe said, "Wherever technology reaches its real fulfillment, it transcends itself into architecture." Everything hinges, of course, on what we mean by *real* fulfillment. This is further complicated by the limitless structural possibilities now open to mankind. "Man's structural imagination has been liberated completely," Pier Luigi Nervi has said. It is now possible to construct buildings and cities that bear resemblance to no others in history, and therefore require a new theory of planning and design.

The need for a radical urban aesthetic, uniting the art of architecture and planning with the science of building, is tragically urgent. Unfortunately, however, as the cities race chaotically towards the twenty-first century, much urban theory—to say nothing of urban practice—adheres timidly to the pre-industrial aesthetics of the nineteenth and early twentieth centuries. I have in mind the new urban sentimentality of Jane Jacobs and others, who are really arguing for a spruced-up status quo. But it is the status quo that has landed us in our present mess, and under the steady pressure of population increase the mess is rapidly growing worse. (I should qualify that: in certain fundamental respects such as public health, for example, our cities are the finest in history, but our potentialities are also the highest in history and we have not come close to attaining them.) In fact, our cities—virtually all large cities, including those subject to the authoritarian regulation of the Soviet Union—are out of control, proliferating wildly. The first step toward bringing them under control must be an analysis of those forces which are now blindly and negatively determining their fate. The next step will be to find rational means of checking those forces and, wherever possible, turning them in a more positive direction.

In this country, particularly, we must devise new controls for land use, which is now anarchic. This means the end of free enterprise in real-estate speculation, but considering the already formidable intrusion of the federal government in the land development business (through guaranteed mortgages, for example, and urban renewal legislation), we no longer have a laissez-faire economy anyway, but a mixed economy.

We must have national and regional, as well as local, planning, and it must be logically synthesized with the development of new sources of energy and new systems of transportation. Water, both as a natural resource and as a source of energy, is fundamental to the life of every individual, but nowhere in the United States is water policy truly consistent with orderly urban development. TVA was a good start, but we have retrogressed in this respect since New Deal days, and in California today we have an extravagant water-development program that I believe will cause more problems than it will solve. I would say that we must plan the environment humanely from the wilderness area in the High Sierras to the street in Los Angeles. If we do, the dividends in human happiness and creativity will be incalculable.

The Choices Ahead

Lewis Mumford

* * * *

Ever since the eighteen-thirties the effects of bad urban planning and bad housing in accentuating all the miseries of inhuman economic exploitation have been recognized. But the attempts to cope with these evils, even in the provision of elementary sanitary facilities, were feeble, superficial, halting, maddeningly slow; and this was especially true in the more congested urban centers, whose very congestion raised land values, and made the vilest slum tenements far more profitable to their landlords than decent houses for the middle classes.

In 1925, when the Regional Planning number of the 'Survey Graphic' was published, only a few people realized that there was something fundamentally wrong with the quality of life in our 'great' and growing American cities, and that far bolder measures than any so far taken were necessary, if these cities were to remain socially well-balanced and attractive places to work and live in. What then seemed to many people healthy evidences of buzzing social activity and economic dynamism were too often, like kiting land values and congested streets, symptoms of social malfunction or organic defects in planning. Most of the evils now so portentously evident in urban communities today were already visible half a century ago—chronic poverty, blighted areas, filthy slums, gangsterism, race riots, police corruption and brutality (the

Originally published as the Postscript to *The Urban Prospect*, by Lewis Mumford (New York, 1968), pp. 227-255.

'third degree'), and a persistent deficiency in medical, social, and educational services.

But the chief proof that something was radically at fault with the whole pattern of life in our metropolitan centers is that those who could afford to leave the city were deserting it—indeed, they had begun to desert long before, seeking in the residential suburb, with its pleasant gardens, its nearby woods and fields, its quiet and safe residential quarters, its neighborly social life, qualities that were steadily disappearing in the more affluent metropolises. And instead of taking this desertion as an instruction to rehabilitate the central city, the leaders of urban society took it as an invitation to invest profitably in multiplying the means of escape, first by railroads, subways, and trolleys, then by motorcars, bridges, tunnels. Automatic congestion was counterbalanced by an equally automatic decongestion and dispersal; and between them, the notion of the city as a socially concentrated, varied, and stimulating and rewarding human environment vanished.

As living conditions worsened in the overcrowded central districts, the area of suburban dispersal widened, until the overflow of one metropolis mingled with the overflow of another metropolis to form the disorganized mass of formless, low-grade urban tissue that is now nicknamed Megalopolis. Like the suburban flight itself, this megalopolitan conglomeration has been treated, often by urban sociologists who should know better, as a recent phenomenon brought about by equally recent technological developments. But Patrick Geddes identified a similar random massing of sub-urban populations more than half a century ago on the British coal fields, and called it, with nicer accuracy, a conurbation—though it turns out that anti-city would be a still better name for it. Observers who now regard this urbanoid massing as the new form of the city, or praise it as a more complex, though unplannable and uncontrollable, substitute for the city, demonstrate that they have never grasped what the historic functions and purposes of the city actually are.

The ultimate mode of this physical dispersal was presciently foretold by H. G. Wells at the beginning of this century in his 'Anticipations of the Reaction of Mechanical and Scientific Progress upon Human Life': but he had, unfortunately, no premonition of the kind of social disintegration that it would bring about. These results are taking form under our eyes, with the result that the residential quarters of our 'great American cities' now tend to separate out into two kinds of ghetto: an upper-class ghetto of high-rise apartments designed, with or without governmental assistance, as status symbols for the super-affluent, and another, lower-class ghetto, scarcely distinguishable from the first on the exterior, for the lowest-income groups. The latter is the home of the new urban proletariat, mainly Negro and Puerto Rican, seeking to escape even worse conditions in San Juan or the Deep South. Those who do not qualify for either ghetto now swell the mass migration to suburbia.

When one translates into concrete terms the current talk about the increasing urbanization of the United States today, one must understand that sociologists are speaking loosely of people who are, in fact, disurbanized, who no longer live in cities, or enjoy, except as visitors or part-time occupants, the concentrated social advantages of the city: the face-to-face meetings, the cultural mixtures, the human challenges. For the growing majority of the non-agricultural population of the United States now lives for better or worse in suburbia: indeed, even many rural areas, where farming is still practiced, are in social content suburban. Meanwhile those who hold fast as residents in the big urban centers—or even in small towns that harbor resentful racial minorities—do so at the peril of their lives. No Berlin walls separate the Gilded Ghetto from the tarnished, oxidized ghettos that spread around them—except in ominously prophetic enclosures like Stuyvesant Town.* But even daylight is not a safeguard against robbery, rape, and murder, as any metropolitan taxi driver will testify.

Yet again we are facing an old situation, which has only been aggravated by the invasion of this new proletariat. The molten human forces that have erupted in a lava flow of violence in our American cities during the last decade have, as in a volcano, manifested themselves earlier in repeated seethings and rumblings. The possibility of such outbreaks of violence was clearly foreseen by Ralph Waldo Emerson more than a century ago. In a passage whose explanatory context the editor of Emerson's 'Journals' unfortunately deleted, he observed: "It is a great step from the thought to the expression of the thought in action. . . . If the wishes of the lowest classes that suffer in these long streets should execute themselves, who can doubt that the city would topple in ruins." Those wishes have now made themselves known from Newark to Detroit, from Boston to Los Angeles.

Even in earlier periods when the repressive forces of law and order seem to have been sufficiently self-confident and ruthless to check any outbreak, the possibility of mass violence lurked under the surface: indeed, one of Louis XIV's acknowledged reasons for building his palace in suburban Versailles was to escape any possible attack by a hostile Paris mob. Catherine Bauer (Wurster), in her pioneer work on 'Modern Housing,' went so far as to suggest that the various efforts to improve the housing of the British working classes in the second part of the nine-

*See Chapter Thirteen in [Lewis Mumford,] 'From the Ground Up,' written in 1948. New York: Harcourt, Brace & World, Inc., 1956.

teenth century were due in part to fear of urban uprisings, a fear awakened by Chartist and Trades Union demonstrations.

Both the grievances and the remedies, then, have a long history: but the grievances were grudgingly acknowledged and half-heartedly corrected, and the remedies were, alas! in part the result of a faulty diagnosis. So the threat of further terror and violence remains—indeed, has been magnified in this generation by the widespread acceptance of methods of terror and violence, in so-called democracies as well as in openly totalitarian regimes, as a 'normal' practice of our civilization, along with bureaucratic regimentation and military compulsion.

Meanwhile, during the last three decades, the racial composition of American cities has changed. Into the great vacuum produced by the suburban exodus has rushed a new army of 'internal immigrants.' Faced with this influx of two depressed minorities, ill-educated, impoverished, usually untrained for work except in agriculture, tens of thousands unable to speak English, American municipalities experienced, in even more acute form, the same difficulties that mass migration from Europe had caused between 1870 and 1920. Though the United States Congress had belatedly sought to ease these difficulties by limiting the number of foreign immigrants admitted in any one year, no attempt was made to direct this internal migration, or limit it to proportions that could be absorbed and assimilated in any one community—still less to spread it over many communities by providing jobs, housing, schools.

The problems raised by these newcomers would have been difficult to handle even if the municipalities concerned had not themselves already been hopelessly in arrears in providing out of their own budgets the necessary schools and hospitals, to say nothing of new housing, for the population already established. Even if state and federal funds had been available in sufficient amounts to provide both housing and rent subsidies—and under the terms of our Cold War economy they were not available—the very volume of this sudden influx would have condemned most of the newcomers to the same verminous, insanitary, congested quarters that, in cities like New York and Boston, they were forced to occupy.

With respect to basic deficiencies—lack of light, air, space, privacy, sanitary services, schools—the grievances of the minorities, both new migrants and those long-established, were well justified. But the slowness of municipal authorities in coming to terms with these grievances only reflected an earlier unreadiness to take any measures for improving the city that did not win the approval of real-estate operators, banks, and insurance companies. What was different between this situation and that which had existed in the eighteen-nineties is that the new immigrants had higher expectations and made new demands.

. . . Though Thomas Jefferson's fears for the physical and moral health of his country if its predominantly rural culture became urbanized and industrialized have long been justified by irrefutable statistical evidence, they are still too often treated by historians as a pathetic bucolic prejudice. Yet the mere increase in urbanization has in fact automatically increased the incidence of pathological social phenomena, as compared with still rural areas. Today's manifestations of hatred, fear, despair, and malevolent violence among the depressed racial and cultural minorities only exhibit in more virulent form the normal pathologies of everyday urban life.

The reason for my silence was neither ignorance nor indifference: I had already dealt with these evidences of urban disintegration in three earlier books. As a student of cities, I knew that all these pathogenic factors had been present in American cities for the last hundred years, along with the inhuman physical congestion and the economic exploitation that have helped to incubate them. Indeed, the entire 'cities movement'—urban sanitation, community centers, parks and playgrounds, municipal health services, publicly aided housing for the lower-income groups—was in large part a response to these degrading conditions. There is nothing new in the present urban situation except the startled public realization that these evils are still with us, on a vaster scale, in a more difficult and desperate form, than ever before.

So far from underrating the human costs of metropolitan congestion and disorganization as merely surface blemishes on the otherwise healthy face of urban life, I had persistently pointed out that they could not be treated by purely temporizing local remedies, since they were symptomatic of deeper organic defects in our civilization. Until now, this point of view has been dismissed as 'unrealistic' or 'pessimistic,' even 'apocalyptic,' by those planners, administrators, and social service workers who sought only to achieve such piecemeal urban improvements as were acceptable and feasible without any critical assessment and renovation of current institutions.

This refusal to look any deeper into the causes of urban deterioration, at a time when the vast surplus of energy, wealth, and knowledge available should have produced a marked improvement of urban life, came out clearly in a common reaction to my book 'The Culture of Cities,' when it was published in 1938. This study of city development was, on the whole, sufficiently hopeful and constructive to be well received. But one section was singled out by certain critics as a dark subjective fantasy, inapplicable to contemporary urban culture. The offending section, 'A Brief Outline of Hell,' was one in which I had summed up the current disintegration of urban life and the probable fate of the city, *if these tendencies continued*. This summary was the restatement of a diagram devised by my old master, Patrick Geddes.

In this diagram Geddes traced the city's evolution through an upward curve, beginning with the polis and reaching a climax in the metropolis, or mother city: then through a downward curve, from megalopolis, handicapped by its own overgrowth, to parasitopolis and pathopolis, till it reached a terminal point: Necropolis, the city of the dead. Those who were eager to discredit Geddes' historic scheme apparently never read to the end of my chapter, where I dealt with 'Possibilities of Renewal' and 'Signs of Salvage.' For had they done so they could not so easily have charged me with holding that Geddes' purely theoretic terminal stage of megalopolitan overgrowth was either necessary or inevitable, still less irreparable.

On the contrary, I had pointed out that cities, not being biological organisms, have often shown signs of senile decrepitude at an early stage, or had undergone processes of renewal at a late moment of their existence, and thus got a new lease on life. And, so far from denying the value of large urban concentrations, I had said that as many as thirty great metropolitan centers might prove necessary to serve as a medium for world intercourse, and as containers of world culture. But the fact that I was aware of the pathological conditions undermining urban life caused many critics to regard me, by some quaint logic of their own, as a sworn enemy of the city.

At the end of 'The Culture of Cities' I had written, with a confidence that had somehow survived the First World War and the economic depression of the thirties: "Already, in the architecture and layout of the new community, one sees the knowledge and discipline that the machine has provided turned to more vital conquests, more human consummations. Already, in imagination and plan, we have transcended the sinister limitations of the existing metropolitan environment. We have much to unbuild, and much more to build: but the foundations are ready. . . ."

The rhetoric now sounds hollow, yet what it suggested seemed possible, even at that late moment. But the Second World War blasted these naive hopes. At the end of that war, instead of laying the foundations for a cooperative civilization, the citizens of the United States put themselves by passive consent in the hands of a "military-scientific-industrial elite," to use President Eisenhower's accurate characterization. By imposing a permanent state of war this 'elite' constructed a vast megamachine, extravagantly supporting and inflating with public funds an assemblage of private corporate mega-machines, operating on the same principles, pursuing the same ends.

This new strategy, based on fantasies of absolute power and total control, placed the mass production of extermination-weapons above human welfare, and so laid the foundations, not for a life economy, but for an anti-life economy, every part of which is elaborately oriented, as in the Egypt of the Pyramid Age, toward death. Witness a regime that spends fifty-seven per cent of its budget every year for military purposes, and has only six per cent available for education, health, and other social services.

But if my hopes for effective urban renovation in America were soon buried, my grimmest apprehensions about the urban future came true more swiftly than I could have anticipated. Only two years after 'The Culture of Cities' came out, the central area of one city after another in Europe was reduced to rubble by aerial bombardments: first, Warsaw and Rotterdam, then London and Berlin, then minor cities in an ever-spreading carnage. Not surprisingly, the harried survivors of this destruction and massacre, as first instituted by the Nazis, did not find my analysis unduly pessimistic: Necropolis lay all around them. Though my work may have had little visible influence in the United States, the Nazi Luftwaffe and its later Allied Air Force imitators had at least given authority to my most ominous predictions, though the invisible moral debacle proved worse—and more permanent—than the visible physical destruction.

Since all my thought about the city had been toward laying the social foundations for urban rebuilding on a regional scale in both old cities and new communities, by stimulating the regenerative and constructive processes already active in our civilization, those who had followed my work were not unprepared for this challenge. This explains why in Europe 'The Culture of Cities' had a far-reaching influence out of proportion, perhaps, to its immediate usefulness; for it was not merely eagerly studied and discussed in England, even while the bombs were dropping, but was used, I have been told, in the Underground architectural schools set up by planners in Poland, the Netherlands, and Greece, to teach the rising generation of planners a new conception of urban development. This situation, at once so menacing and yet so promising, was an incentive to further thinking on my own part.

In 1945, accordingly, in a critique of Abercrombie and Forshaw's plan for Greater London, I outlined specifically the further steps that should be taken—apart from the needed building of New Towns—to prevent the further congestion of London, and to make possible its rebuilding on a more human scale.* One of these steps was the local decentralization of governmental and business offices into the constituent Boroughs of London, in order to lessen the daily commuter traffic to Central London and restore the metropolis itself as a place of residence, with amenities equal or superior to those of any suburb, and with greater facilities for human association, unpenalized by time-wasting, energy-depleting travel.

*See [Lewis Mumford,] 'City Development: Studies in Disintegration and Renewal.' New York: Harcourt, Brace & World, Inc., 1945.

These specific proposals—the building of a series of New Towns, with the removal of suitable industries and bureaux from congested areas to relatively empty ones, the planning of neighborhoods to facilitate family life and autonomous communal acitivity, the creation of regional authorities to direct the work of urban development over a wider area—were, in fact, all carried on vigorously in Britain after 1947, with the exception of the last item; and even this necessary extension of urban authority, from the metropolis to the region, is now under active discussion. Whatever further initiatives and modifications may be needed, these measures have all proved practical; and in the case of the most disputed and disparaged proposal, that for a large-scale building of New Towns to provide both industrial and social advantages that no congested metropolis can offer, these towns have proved immensely successful—so successful, in fact, that canny speculators even attempted a 'take-over' of the oldest New Town, Letchworth, lured by the prospective increase in values. However modest my own contributions have been to this program, they at least antedated the postwar legislation and building.

But I have an arresting objective reason rather than a personal one for dwelling on these details. And this is to point out that despite Britain's immense constructive achievements in housing and planning and the industrial replenishment of underdeveloped areas, the same general disintegration and demoralization that has been going on in other parts of Western civilization has gone on there. This can no longer be attributed to postwar exhaustion. Three centuries of brutal exploitation, enslavement, destruction, and extermination have left their mark on civilized society. In England now, no less than in the United States, the same marks of urban disintegration have nevertheless appeared in massive quantities—police corruption, marital promiscuity, random reproduction, overt racial and class antagonism, narcotic addiction, cultivated sadism, defiant criminality. The cult of anti-life, symbolically prefigured in much of the *avant-garde* art and music and drama of our time, is now spreading actively into every part of megalopolitan routine. Patholopolis and Parasitopolis, in fact, are fast establishing themselves as normal forms of the city, or, rather, as negative heavens: ideal environments for the psychotic, the criminal, the feckless, and the demoralized. The terminal stage in city development would seem nearer than ever.

Now, in all societies, the upbuilding and the breaking-down processes go on side by side, as they do in living organisms. As long as the constructive processes are dominant the organism survives, and to the extent that it has a margin of free energy and maintains its powers of self-direction and self-replication, the organism may flourish. What makes the present situation so singular and so threatening, is that the extra energies available, when not claimed by the production of lethal weapons and space rockets, are absorbed by the centralized bureaucratic and technological processes that are scattering the specialized parts of the city over the landscape. These individual urban groups and communities no longer have effective control over their own destinies. As a result, if anything goes wrong locally, the defective part, so to say, can no longer be repaired on the spot, but must be 'sent back to the factory.'

These facts have convinced me, and I think should convince any unbiassed observer, that the underlying causes for the recurrent outbreaks of violence among the disturbed minorities are not to be found solely in the sordid physical conditions of the cities themselves. While the recent demonstrations and revolts are partly accountable as a long-delayed reaction to poverty, slum housing, unemployment, social discrimination, police animosity, and segregation, the cities that have taken the most vigorous measures to deal with these evils, like Detroit and New Haven, have proved no more immune to attack than those that have been inert and indifferent. So, though the continued effort to turn the city into a comely, life-fulfilling environment is still one of the great collective tasks of our day, it is not a panacea. Such efforts will enhance the goodness of the city's goods; but they will not abate the evil of its real evils, since the latter are not under local control, nor have they only a local origin.

Those who now impatiently demand, or confidently prescribe, a heavy national investment in good housing or a 'model cities program,' as an antidote for demonstrative mass violence or as a curb to juvenile delinquency and adult criminality have not looked carefully enough at the evidence. If juvenile delinquency, for example, were mainly the result of poverty and alienation, why should it break out equally in spacious upper-class, white American suburbs? Certainly the common denominator here is not a bad physical environment. To ask the legislator or the planner to apply such immediate remedial measures to restore order is to ask for quackery. It is not just the city but the whole body politic that demands our attention. The advertiser's mirage of the Affluent Society may tease and torment the depressed minorities that are denied a full share of this affluence, but the reality itself appals the overfed, overstimulated, overcoddled young who are bored by its smooth lubricity.

These outbreaks are but local incidents in the vast eruptions and lava flows of collective violence that mark the last half century as the most violent age in history, with a record of wholesale destruction and merciless extermination that makes the most savage conquests of the

Assyrians, the Tartars, and the Aztecs seem the work of diffident amateurs. What has been happening in our cities can be neither understood nor controlled except in the light of this larger example of insensate destruction. The progressive technology that the Victorian exponents of industrialism looked upon as a certain means of assuring peace and plenty has been increasingly corrupted by its commitment to organized nihilism and aggression. Its greatest achievements—nuclear bombs, computers, radar, rockets, supersonic planes—are all by-products of war. Constant indoctrination in violence is the main office of our ubiquitous agents of mass communication and mass education. To believe that a single organ of the body politic, the city, can be cured of this disease while the same deadly cells flow through the entire bloodstream is to betray an ignorance of elementary physiology.

One thing should at least be obvious by now. Neither the past diagnoses of urban defects nor the positive regimens offered for urban health have proved competent or effective. So, though the kind of constructive planning I have advocated in these chapters is still viable, and indeed more urgent now than ever, it would be foolish to put forth these proposals as a means of averting future gang rumbles, 'race riots,' or Negro-Puerto Rican revolts. That situation has another dimension.

The need to make a fresh start on a more human basis is imperative, as many of the younger generation—at least those who have not 'dropped out'—plainly realize. But this fresh start must be made at many different points, and what is done in one area must, from the beginning, be coordinated with many similar efforts to restore effective human initiative and human goals. Perhaps the surest way of abating urban violence in the United States would be to set a striking national example of moral continence by the voluntary withdrawal of American forces from Vietnam. This could be more effective in reversing the currents of violence than even the dedication of the vast sums of money and energy wasted in military adventures to the imaginative replenishment of American life. But it might take a whole generation before even such an act of national self-discipline and moral atonement would make a positive contribution to comity and order.

At all events, to suppose that a fresh start can be made merely by pouring millions of dollars into the same public housing and urban renewal projects that have already proved so futile would be to nourish further illusions. This is like prescribing massive doses of penicillin to a patient in the terminal stages of a chronic disease—though, at an earlier moment, diet and surgery might have cured him. No quick miraculous recovery can now be hoped for; or, rather, the one conceivable miracle that might yet occur is that a sufficient number of people should recognize that every part of our life must be overhauled, including "the technology of Megalopolis" and the supporting ideology of an affluent society under an ever-expanding economy.

This larger theme is not one that I can handle even sketchily in a postscript. Many contemporary thinkers have at least made a beginning in diagnosing our present situation, from Spengler, Toynbee, and Schweitzer onward; and I have made an extensive contribution in a series of books, most recently in 'The City in History' and 'The Myth of the Machine.' In this final comment I shall only pose some of the difficult immediate problems that neither the dissident minorities, justifiably outraged and impatient, nor the once blindly complacent majority have so far been willing to face.

. . . *Eventually far vaster sums than are now being proposed will be needed: at least the equivalent of our present vast annual military expenditure.* When I said "Go Slow" I meant "Do not hastily pour tens of billions into any national program until the mistakes that had been made during the last quarter century have been analyzed and corrected, until new procedures and new plans have been worked out, until new agencies on a regional basis are created—indeed, until the whole program is completely rewritten, and more humane minds in architecture, planning, and administration are installed as leaders." Such leaders must be capable of both learning from and patiently teaching those groups for whom these projects are intended. Technological and architectural exhibitionists, more interested in abstract models and structures than in concrete human realities, have no place in such a program.

* * * *

What made me specially apprehensive that we might, in desperation, be on the point of committing more crippling errors than in the past was Senator Ribicoff's published suggestion that a more humane type of housing, based on the neighborhood, could be done quickly and efficiently by calling upon "the technology of Megalopolis," for this was the very technology whose ruthless dynamism had already done so much to break down the coherent structure of the historic city. This technology is based upon accommodating human requirements to the demands of automation, mass production, maximum financial profit, and remote control: it would dismiss any human needs, desires, appetites, and sentiments that do not conform to its system of production as rigorously as it would dismiss an astronaut's normal freedoms in the design of a space capsule.

In a small way, as I pointed out, "the technology of Megalopolis" had already exhibited its characteristic absence of human dimensions in our current urban renewal projects. These projects were designed to satisfy the needs of the real-estate operator, the building contractor, the municipal administrator, the governmental bureaucracy—but without any respect for the traditions, the desires, or the hopes of the displaced slum dwellers, only a small part of whose numbers, even when qualified, have in fact consented to occupy the new buildings.

On this radical failure, the evidence was already at hand: in the new quarters sunshine, fresh air, modern plumbing, modest rents had cancelled out old physical defects: but this had not satisfied more basic human desires, eased anxieties, torments, resentments, or fostered self-respect, self-control, and autonomous action. Such residual autonomy as could be recaptured came mainly through vice and crime.

Sterile and humanly inadequate though current public housing has been, its failure to do justice to human needs would be nothing to what "the technology of Megalopolis" would achieve, under the aegis of such typical corporations as I satirically nicknamed 'General Space Housing, Incorporated.' The coming generation of architects, already conditioned to the neatly programmed world of computers, insulated from ecological realities, organic and human, has far worse designs in store. The draughting boards in architectural schools are nowadays full of anti-urban designs for nonliving: projects for heaping fifty thousand people, or even a million, into a single structure, now in the air, now submerged in water, now under a geodesic dome: highly ingenious structures, each as original as a space capsule, and as unfitted for permanent human habitation.

The one factor left out of these designs is precisely the factor that would reveal what is wrong with these go-go structures: the faintest concern for what kind of houses, neighborhoods, and cities people actually want to live in. There is no single answer to this problem, because even durable wants change socially and esthetically from generation to generation. But the one answer that is no answer at all is to ignore these wants in order to come up with a uniform solution that completely satisfies the criteria of systems analysis, mass production, and automation. To reduce the human factors to those a computer can handle is to eliminate the better half of life. The current catch phrase 'instant cities' is a perfect example of these sterile anti-historic conceptions put forward by under-dimensioned minds.

Whenever a poll has been taken to find out what kind of housing people really want, it turns out that an overwhelming majority vote for the single-family house. In a sense, the results of such polls have been anticipated in the wholesale exodus to the suburb; but even in a country like France, where city people have long been addicted to flats, an overwhelming majority in a recent poll—eighty per cent—voted for the single-family house.* Though the financial success of a series of Levittowns confirms the wisdom of heeding this demand, the needless esthetic dreariness and social monotony of these particular suburban enclaves—*pace* Herbert Gans!—show that even more humane intentions in mass production succumb to the limitations of the method itself.

To suppose that the abstract intelligence that proved so brilliant in developing nuclear reactors and supersonic planes will be equally successful in finding technological solutions that could be applied wholesale to the rebuilding of our cities is to misconceive the nature of the whole problem. There is little prospect of overcoming the errors that have been made, first in building and extending our cities, and now in repairing and reconstructing their more outmoded or outlived structures, unless we put the human condition first. This means that we must be ready, before legislating or planning, to revise the obsolete premises, financial, political, architectural, and technological—not least, technological—upon which our constructive efforts have so far been based. The kind of thinking needed is too important to be left to experts and specialists. In this situation we are all amateurs, for our life experience is more important than any technical expertise. Hence a solitary walk through an urban slum may provide more valuable insight than millions of Foundation dollars spent for statistically impeccable research.

The shudder of apprehension that followed the recent outbreaks of mass violence in many once-orderly American cities has perhaps produced a body of citizens at last willing to face and to cope with the cumulative anti-social realities. Since the ameliorative measures so far followed have proved ineffective, there is room for a more comprehensive view of the urban situation, and for more far-reaching constructive proposals, based on deeper insight into our whole civilization. We are faced not only with the task of physical renovation, hygienic or esthetic: we have the burden of offering a valid alternative to a kind of existence that the neglected and the deprived are no longer reconciled to, because they suspect that something better is possible, indeed, almost within their reach, but arbitrarily withheld from them. On these essential human concerns, as I remarked to the Ribicoff Committee on this matter, Studs Terkel's 'Division Street' or Oscar Lewis' 'La Vida' have more to teach us than our multiplying Institutes for Urban Research.

Now I have no special competence to analyze the immediate social factors that have brought on the current eruptions of violence which

* See the three volumes entitled '*Les Pavillonaires*.' Paris: Centre de Recherche d'Urbanisme, 1966.

make the task of urban improvement so pressing. Here I must trust the abundant testimony of those who have been in close daily contact with the minority groups concerned. But I know enough of the history of modern metropolises, like my own city of New York, to realize that poor sanitation, congested, bug-ridden, rat-infested dwellings, and desperate poverty are not in themselves sufficient to promote more than comparatively minor hoodlumism and gangsterism, since all these conditions had been chronic for almost a century among a diversity of white immigrant populations without provoking comparable riots.

To go deeper into this immediate situation we must, I suggest, distinguish between three aspects, only one of which is open to immediate rectification. We must first separate out the problems that are soluble with the means we have at hand: this includes such immediate measures as vermin control, improved garbage collection, cheap public transportation, new schools and hospitals and health clinics. Second, those that require a new approach, new agencies, new methods, whose assemblage will require time, even though the earliest possible action is urgent. And finally, there are those that require a reorientation in the purposes and ultimate ideals of our whole civilization—solutions that hinge on a change of mind, as far-reaching as that which characterized the change from the medieval religious mind to the modern scientific mind. Ultimately, the success of the first two changes will hinge upon this larger—and, necessarily, later—transformation. So, far from looking to a scientifically oriented technology to solve our problems, we must realize that this highly sophisticated dehumanized technology itself now produces some of our most vexatious problems, including the unemployment of the unskilled.

Let me touch on the hardest aspect first, for though the goal indicated is remote, a beginning should be made at once. In the most general terms, this basic problem is the control of power, quantification, automatism, aimless dynamism. That problem has become acute in our age, because scientific technology has colossally magnified the amount of energy that advanced industrial societies command. But even more, it has become difficult because in our overreliance upon purely intellectual enlightenment we have allowed earlier systems of moral, political, and social control to break down, and have transferred systematic discipline and order to the very corporate organizations that must be brought once more under human direction, if they are to pursue human ends.

Once the traditional system of moral restraints and personal inhibitions has dissolved in any society, as completely as has happened during the last half century throughout the Western world, the warfare of each against all, which Thomas Hobbes falsely pictured as the original state of primitive man, becomes more than a theoretic possibility: it has, in fact, become a demonstrable reality. And unfortunately, the very institution that Hobbes relied upon to put down this internecine strife, the Leviathan state, is now the chief offender in flouting law and order, in extending the sphere of violence and magnifying all the possibilities of destruction and extermination. In effect, the policeman is the chief criminal, and his bad example has proved infectious.

There is not a single human problem posed in our cities, as between White Power and Black Power, that was not prefigured in the last three centuries of conquest, colonization, enslavement, exploitation, and extermination; and there is not a difficulty faced by the United Nations, seeking to achieve a balance between tribalism and universalism, between nationalism and cosmopolitanism, that will not have to be worked out in the smallest nieghborhood.

The forces that have violated the elemental moralities and that now threaten all life on this planet will not be easily or quickly brought under control. But in so far as command of these menacing forces means imposing salutary inhibitions, restraints, prudences, it is open to every sane, responsible person to make a beginning in his own life. Only those who have lost respect for the principle of autonomy may either 'go with' the forces of disintegration or express their disillusioned dissent merely by 'dropping out.'

* * * *

Unfortunately, even some of the urban problems that would seem to be immediately soluble, once we accepted the high price of a solution, are not quite so simple as reformers have hitherto believed. To begin with, consider desegregation. The mass migration of Negroes and Puerto Ricans into Northern communities has turned once-diffused minorities into concentrated metropolitan enclaves that will soon, if present tendencies continue, constitute a hapless proletarian majority. No open housing or school busing can overcome the isolation and resultant self-segregation that sheer numbers have produced. Before any urban renewal program can be instituted for the benefit of given racial or cultural groups, the first question that must be answered, by the minorities themselves, is whether they are willing to move out of their present neighborhoods, even if this means scattering widely in a mixed community, and losing some of their present identity and cohesion.

If they choose to remain in numbers where they are, they commit themselves to continued segregation: not merely to segregation but also to congestion, and, along with congestion, to insufficient recreation areas and overcrowded health and hospital services, too. But if they choose to move far enough away to invite the provision for good housing, new industrial or agricultural opportunities, and stable neighborhood facilities, they will become a part of New Towns, suburbs, or growing rural communities; and they will, like any other newcomers, perhaps need a generation before they are fully integrated, no matter how faithfully their legal status as citizens is secured.

This decision cannot be made in local City Halls, still less in Washington; for only those concerned have the right to make it, after the way has been sufficiently opened by experimental planning and building to make a genuine choice possible. Yet no intelligent program of urban renewal can be framed until this alternative has been built into the program itself. Only one thing can be predicted: if the immigration to big cities and the metropolitan birth rate continues at the past high levels, there will be no alternative to organizing dispersal and relocation, both regionally and nationally, into smaller communities. Fortunately, a rational program for resettlement—what was called in the Preface 'the Fourth Migration'—is still an open option, more open now than ever, because so many industries and business organizations have been, during the last two decades, sporadically moving away from the metropolis.

But the underlying human factors are still too delicate and uncertain to admit quick decisions. The policy of dispersal now quietly favored by educated middle-class Negroes in professions and businesses, thanks partly to Morris Milgrim's initiative in providing open suburban housing, would leave the metropolitan ghettos without leadership; and so, worse off socially than before. Dispersal would also have the effect of undermining the sub-culture that has developed in Harlem and other major metropolitan centers: yet this sub-culture, through its expression in music, the dance, and the theater, is one of the chief sources of the Negro's and the Puerto Rican's individuality and self-respect. (Certainly, something was lost to the once-thriving Yiddish sub-culture by its voluntary removal to improved living quarters away from the Lower East Side.) By now there is an activist Negro minority—how large is anyone's guess—that would resist such assimilation. Neither choice is clear and easy: so both must be kept open.

But if the slums and the blighted areas where the minorities now live were to be rehabilitated for the existing overcrowded population, this would mean continuing to build superslums, whether on the open high-rise pattern favored by municipal authorities, or the dense housing on crowded lots, without provisions for sunlight, open air, or visual amenity, favored by Jane Jacobs, in which streets would remain the chief play areas, though filled, as now, with dangerous traffic. Neither the frying pan nor the fire is attractive; for housing designed at three hundred to four hundred people to the acre—to say nothing of the greater number some favor—is not conducive to health, neighborly cooperation, or adequate child care. The slum dwellers' justifiable resentment against arbitrary uprooting and his unwillingness to return to the kind of inhuman high-rise apartments offered has now been fully demonstrated; and to go on building in this fashion would be foolish.

The core of any adequate neighborhood housing program should be, above all, the provision for the health, security, education, and adult care of young children; and except perhaps in health, all high-rise projects are, by their very scale and impersonality, an alien and even hostile environment for the young, since, apart from organized playground games, it leaves the majority of children such little scope for their own activities. In these new quarters, even the mildest outbreak of juvenile adventure or wanton mischief becomes all too quickly labelled as juvenile delinquency—on which terms, Robert Frost once confessed to me, he probably would have spent his own boyhood in San Francisco in a reformatory.

Lacking both normal parental disciplines and normal outlets for defiance of adults, something worse now takes place. One of the most sinister features of the recent urban riots has been the presence of roaming bands of children, armed with bottles and stones, taunting and defying the police, smashing windows and looting stores. But this was only an intensification of the window-breakings, knifings, and murders that have for the past twenty years characterized "the spirit of youth in the city streets."

As I pointed out earlier, juvenile delinquency is not confined to a depressed minority living in slums: it is also an upper- or middle-class white, suburban phenomenon. But in both cases it seems to point to two underlying conditions: an idle, empty, purposeless existence, and a total breakdown of parental guidance and communal discipline. In both groups we find, among the younger adults, that marital promiscuity and parental irresponsibility have undermined the basic unit of all stable societies—the family.

This family disintegration can only in part be attributed to bad housing. Unfortunately, it has been worsened by what was, by intention, a humane achievement in legislation: the provision of welfare relief to mothers solely responsible for their children's support. This legislation turns out, in the case of husbandless mothers, too often to be a subsidy to sexual irresponsibility and an invitation to chronic idleness. The demoralizing effects of this remedy come out in the disturbing, if perhaps apocryphal, story of the little girl brought up under such conditions who, when asked what work she intended to do when she grew up, replied that she wanted to draw. Inquiry revealed that what she wanted to 'draw' was not pictures, but a welfare check, as her mother did.

Obviously, the high rate of unemployment among Negroes and Puerto Ricans, and the lower wages and poorer conditions offered needier groups, both colored and white, discourages stable marriages, and perhaps vitiates male parental feeling as well. But to think of correcting this condition solely by rent subsidies, that is, in effect, more welfare checks, or by better physical housing, is to overlook the equal need for active responsive cooperation by those concerned. Promiscuity cannot be legally suppressed: but marital stability and parental prudence could be honorably rewarded, not only by year-round employment, but by family wages to the fathers of families, as in France; financed in the United States as Social Security is financed—with bonuses that would cease after the third child. This lies outside the scope of urban renewal; but no adequate renewal program will be possible until the restoration of the basic family constellation is taken as one of the essential goals of adequate housing.

On this matter, if one can safely accept a recent report, the example of Hong Kong is pertinent, for it would seem to show that moral factors count for more than purely physical ones, as soon as a certain minimum standard of sanitation has been achieved. In that oppressively congested metropolis, high-rise housing has been provided for the lowest-income groups at much higher densities than any housing authority has dared to establish in America. About the best that can be said for these quarters is that they are ratproof, fireproof, sanitary. Since the parents and the older children must work, the very young are habitually locked in their flats all day. On the surface, these grim conditions would seem to intensify all the domestic difficulties and juvenile disturbances that characterize high-rise housing for lower-income groups in America.

But however far from ideal the conditions for family life are in the municipal tenements of Hong Kong, they are partly redeemed, it would seem, by two factors not present in contemporary American communities: one is that the Chinese cult of the family still prevails, with the young conditioned to respect their elders and accept their authority, while the parents shoulder their responsibility; and the other is that, under bitter pressure of necessity, every member of the family, old and young, has daily work to do. Thus the young are demoralized neither by the breakdown of the family nor by the absence of active duties and serious tasks; nor yet are their parents haunted by such dreams of effortless affluence as would make their present poverty harder to bear. Even Hong Kong's sweatshop labor seems less demoralizing than total idleness. To protect the young from overidleness has now become as important as it was once to protect them from overwork. To this end both child labor legislation and trade union regulations need to be judiciously overhauled.

One would have reservations about this Hong Kong report, but for the fact that it is confirmed by earlier American experience. Much the same conditions for stability prevailed in American cities among the older immigrant groups; for they were held together by Old World village loyalties, by family closeness, by religious precept and ritual: such hopes as they cherished for a better future were based upon their own foresight, thrift, and self-education. The physical conditions of life in the nineteenth-century slums were often as bad as those in Harlem today and much worse than those in Watts: but there were strong moral counterweights that have now been lost through the more general dissolution of human values.

I have said enough to indicate that neither public housing, slum clearance, nor neighborhood rehabilitation, even when done along more human lines than those now in evidence in urban renewal areas, will suffice by themselves to overcome the internal disorders of the city. Those disorders are symptoms of a wider moral breakdown in our whole civilization; and though good planning, like pure water, is essential to urban health, it is not any more than water a prescription for curing disease. Anything worthy to be called urban renewal today must recapture in concrete form many of the values that our affluent, remote-controlled, electronically massaged society has lost. And there is no urban program that one might offer to minority groups that is not just as imperatively applicable to the rest of society. In that sense, there is no Negro problem and no Puerto Rican problem: there is only a human problem.

On this subject we should do well to heed Dr. C. G. Jung's observations about his own life. In his autobiography, 'Memories, Dreams, Reflections,' Jung recalled a difficult period when he was in a psychotic state, at the mercy of his unconscious. What kept him from going completely to pieces was his consciousness that he was an identifiable person, with a family to support, that he was a member of a respected profession, living in a particular house, in an equally familiar and recognizable city, where he had daily duties to perform. By clinging to these reassuring evidences of stability, he was able to resist the internal forces of disintegration.

All of these vital conditions for social continuity and personal integrity have been breaking down in both the central metropolis and its outlying areas; and they have most completely broken down among the lowest-income groups. This unfortunate minority lacks regular work and the self-respect that comes from performing such work: their immediate neighborhood and city have undergone and are still undergoing abrupt structural changes, for both bad and good, that erase their familiar social patterns and destroy their sense of belonging, so that their own selves become so much scattered debris in the larger demolition process. Neither family nor property nor vocational respect nor an earned income nor an identifiable home helps the segregated or displaced minority to resist further internal disintegration.

In analyzing the conditions that saved him from disruption, Jung demonstrated the unique advantage of the historic city over the unstable, incoherent, haphazardly dispersed Megalopolis. In that act, he put his finger likewise on the essential requisites for overcoming the forces that have been disintegrating and dehumanizing both our cities and our civilization.

PART V

POPULATION

Harder Than Granite

It is a pity the shock-waves
Of the present population-explosion must push in here too.
They will certainly within a century
Eat up the old woods I planted and throw down
  my stonework: Only the little tower,
Four-foot-thick-walled and useless may stand for a time.
That and some verses. It is curious that flower-soft verse
Is sometimes harder than granite, tougher than a
  steel cable, more alive than life.

Robinson Jeffers

From *The Beginning and the End and Other Poems,* by Robinson Jeffers (New York, 1963), p. 58.

The Ideal State Described

Aristotle

The first point to be considered is what should be the conditions of the ideal or perfect state; for the perfect state cannot exist without a due supply of the means of life. And therefore we must presuppose many purely imaginary conditions, but nothing impossible. There will be a certain number of citizens, a country in which to place them, and the like. As the weaver or shipbuilder or any other artisan must have the material proper for his work (and in proportion as this is better prepared, so will the result of his art be nobler), so the statesman or legislator must also have the materials suited to him.

First among the materials required by the statesman is population: he will consider what should be the number and character of the citizens, and then what should be the size and character of the country. Most persons think that a state in order to be happy ought to be large; but even if they are right, they have no idea what is a large and what is a small state. For they judge of the size of the city by the number of the inhabitants; whereas they ought to regard, not their number, but their power. A city too, like an individual, has a work to do; and that city which is best adapted to the fulfillment of its work is to be deemed greatest, in the same sense of the word great in which Hippocrates might be called greater, not as a man, but as a physician, than some one else who was taller. And even if we reckon greatness by numbers, we ought not to include everybody, for there must always be in cities a multitude of slaves and sojourners and foreigners; but we should in-

From *The Politics of Aristotle*, translated by B. Jowett (Oxford, 1855), pp. 213-215, 240.

clude those only who are members of the state, and who form an essential part of it. The number of the latter is a proof of the greatness of a city; but a city which produces numerous artisans and comparatively few soldiers cannot be great, for a great city is not to be confounded with a populous one. Moreover, experience shows that a very populous city can rarely, if ever, be well governed; since all cities which have a reputation for good government have a limit of population. We may argue on grounds of reason, and the same result will follow. For law is order, and good law is good order; but a very great multitude cannot be orderly: to introduce order into the unlimited is the work of a divine power—of such a power as holds together the universe. Beauty is realized in number and magnitude, and the state which combines magnitude with good order must necessarily be the most beautiful. To the size of states there is a limit, as there is to other things, plants, animals, implements; for none of these retain their natural power when they are too large or too small, but they either wholly lose their nature, or are spoiled. For example, a ship which is only a span long will not be a ship at all, nor a ship a quarter of a mile long; yet there may be a ship of a certain size, either too large or too small, which will still be a ship, but bad for sailing. In like manner a state when composed of too few is not as a state ought to be, self-sufficing; when of too many, though self-sufficing in all mere necessaries, it is a nation and not a state, being almost incapable of constitutional government. For who can be the general of such a vast multitude, or who the herald, unless we have the voice of a Stentor?

A state then only begins to exist when it has attained a population sufficient for a good life in the political community: it may indeed somewhat exceed this number. But, as I was saying, there must be a limit. What should be the limit will be easily ascertained by experience. For both governors and governed have duties to perform; the special functions of a governor are to command and to judge. But if the citizens of a state are to judge and to distribute offices according to merit, then they must know each other's characters; where they do not possess this knowledge, both the election to offices and the decision of lawsuits will go wrong. When the population is very large they are manifestly settled at haphazard, which clearly ought not to be. Besides, in an over-populous state foreigners . . . will readily acquire the rights of citizens, for who will find them out? Clearly then the best limit of the population of a state is the largest number which suffices for the purposes of life, and can be taken in at a single view. Enough concerning the size of a city.

* * * *

As to the exposure and rearing of children, let there be a law that no deformed child shall live, but where there are too many (for in our state population has a limit), when couples have children in excess, and the state of feeling is averse to the exposure of offspring, let abortion be procured before sense and life have begun; what may or may not be lawfully done in these cases depends on the question of life and sensation.

A Modest Proposal for Preventing the Children of Poor People in Ireland from Being a Burden to their Parents or Country, and for Making Them Beneficial to the Public

Jonathan Swift

It is a melancholy object to those who walk through this great town, or travel in the country, when they see the streets, the roads, and cabin doors crowded with beggars of the female sex, followed by three, four, or six children, all in rags, and importuning every passenger for an alms. These mothers, instead of being able to work for their honest livelihood, are forced to employ all their time in strolling to beg sustenance for their helpless infants; who, as they grow up, either turn thieves, for want of work, or leave their dear native country to fight for the pretender in Spain, or sell themselves to the Barbadoes.

I think it is agreed by all parties, that this prodigious number of children in the arms, or on the backs, or at the heels of their mothers, and frequently of their fathers, is, in the present deplorable state of the kingdom, a very great additional grievance; and therefore whoever could find out a fair, cheap, and easy method of making these children sound useful members of the commonwealth, would deserve so well of the public, as to have his statue set up for a preserver of the nation.

But my intention is very far from being confined to provide only for the children of professed beggars: it is of a much greater extent, and shall take in the whole number of infants at a certain age, who are born

From *The Works of Jonathan Swift*, edited by Walter Scott (Edinburgh, 1814), pp. 454-466. Spelling and punctuation have been modernized.

of parents in effect as little able to support them, as those who demand our charity in the streets.

As to my own part, having turned my thoughts for many years upon this important subject, and maturely weighed the several schemes of our projectors, I have always found them grossly mistaken in their computation. It is true, a child just dropped from its dam may be supported by her milk for a solar year, with little other nourishment: at most not above the value of two shillings, which the mother may certainly get, or the value in scraps, by her lawful occupation of begging; and it is exactly at one year old that I propose to provide for them in such a manner, as, instead of being a charge upon their parents, or the parish, or wanting food and raiment for the rest of their lives, they shall, on the contrary, contribute to the feeding, and partly to the clothing of many thousands.

There is likewise another great advantage in my scheme, that it will prevent those voluntary abortions, and that horrid practice of women murdering their bastard children, alas too frequent among us! sacrificing the poor innocent babes, I doubt more to avoid the expense than the shame, which would move tears and pity in the most savage and inhuman breast.

The number of souls in this kingdom being usually reckoned one million and a half, of these I calculate there may be about two hundred thousand couple whose wives are breeders; from which number I subtract thirty thousand couple, who are able to maintain their own children (although I apprehend there cannot be so many, under the present distresses of the kingdom); but this being granted, there will remain a hundred and seventy thousand breeders. I again subtract fifty thousand, for those women who miscarry, or whose children die by accident or disease within the year. There only remain a hundred and twenty thousand children of poor parents annually born. The question therefore is, How this number shall be reared and provided for? which, as I have already said, under the present situation of affairs, is utterly impossible by all the methods hitherto proposed. For we can neither employ them in handicraft or agriculture; we neither build houses (I mean in the country), nor cultivate land: they can very seldom pick up a livelihood by stealing, till they arrive at six years old, except where they are of towardly parts; although I confess they learn the rudiments much earlier; during which time they can however be properly looked upon only as probationers; as I have been informed by a principal gentleman in the county of Cavan, who protested to me, that he never knew above one or two instances under the age of six, even in a part of the kingdom so renowned for the quickest proficiency in that art.

I am assured by our merchants, that a boy or a girl before twelve years old is no salable commodity; and even when they come to this age they will not yield above three pounds, or three pounds and half-a-crown at most, on the exchange; which cannot turn to account either to the parents or kingdom, the charge of nutriment and rags having been at least four times that value.

I shall now therefore humbly propose my own thoughts, which I hope will not be liable to the least objection.

I have been assured by a very knowing American of my acquaintance in London, that a young healthy child, well nursed, is at a year old a most delicious, nourishing, and wholesome food, whether stewed, roasted, baked, or boiled; and I make no doubt that it will equally serve in a fricassee or a ragout.

I do therefore humbly offer it to public consideration, that of the hundred and twenty thousand children already computed, twenty thousand may be reserved for breed, whereof only one-fourth part to be males; which is more than we allow to sheep, black-cattle, or swine; and my reason is, that these children are seldom the fruits of marriage, a circumstance not much regarded by our savages, therefore one male will be sufficient to serve four females. That the remaining hundred thousand may, at a year old, be offered in sale to the persons of quality and fortune through the kingdom; always advising the mother to let them suck plentifully in the last month, so as to render them plump and fat for a good table. A child will make two dishes at an entertainment for friends; and when the family dines alone, the fore or hind quarter will make a reasonable dish, and seasoned with a little pepper or salt, will be very good boiled on the fourth day, especially in winter.

I have reckoned, upon a medium, that a child just born will weigh twelve pounds, and in a solar year, if tolerably nursed, will increase to twenty-eight pounds.

I grant this food will be somewhat dear, and therefore very proper for landlords, who, as they have already devoured most of the parents, seem to have the best title to the children.

Infants' flesh will be in season throughout the year, but more plentifully in March, and a little before and after: for we are told by a grave author, an eminent French physician, that fish being a prolific diet, there are more children born in Roman catholic countries about nine months after Lent, than at any other season; therefore, reckoning a year after Lent, the markets will be more glutted than usual, because the number of popish infants is at least three to one in this kingdom; and therefore it will have one other collateral advantage, by lessening the number of papists among us.

I have already computed the charge of nursing a beggar's child (in which list I reckon all cottagers, laborers, and four-fifths of the farmers) to be about two shillings per annum, rags included; and I believe no gentleman would repine to give ten shillings for the carcass of a good fat child, which, as I have said, will make four dishes of excellent nutritive meat, when he has only some particular friend or his own family to dine with him. Thus the squire will learn to be a good landlord, and grow popular among his tenants; the mother will have eight shilling neat profit, and be fit for work till she produces another child.

Those who are more thrifty (as I must confess the times require) may flay the carcass; the skin of which, artifically dressed, will make admirable gloves for ladies, and summer-boots for fine gentlemen.

As to our city of Dublin, shambles may be appointed for this purpose in the most convenient parts of it, and butchers we may be assured will not be wanting; although I rather recommend buying the children alive than dressing them hot from the knife, as we do roasting pigs.

A very worthy person, a true lover of his country, and whose virtues I highly esteem, was lately pleased, in discoursing on this matter, to offer a refinement upon my scheme. He said that many gentlemen of this kingdom, having of late destroyed their deer, he conceived that the want of venison might be well supplied by the bodies of young lads and maidens, not exceeding fourteen years of age, nor under twelve; so great a number of both sexes in every county being now ready to starve for want of work and service; and these to be disposed of by their parents if alive, or otherwise by their nearest relations. But with due deference to so excellent a friend, and so deserving a patriot, I cannot be altogether in his sentiments; for as to the males, my American acquaintance assured me, from frequent experience, that their flesh was generally tough and lean, like that of our schoolboys, by continual exercise, and their taste disagreeable; and to fatten them would not answer the charge. Then as to the females, it would, I think, with humble submission, be a loss to the public, because they soon would become breeders themselves: and besides, it is not improbable that some scrupulous people might be apt to censure such a practice, (although indeed very unjustly) as a little bordering upon cruelty; which, I confess, has always been with me the strongest objection against any project, how well soever intended.

But in order to justify my friend, he confessed that this expedient was put into his head by the famous Psalmanazar, a native of the island Formosa, who came from thence to London above twenty years ago; and in conversation told my friend, that in his country, when any young person happened to be put to death, the executioner sold the carcass to persons of quality as a prime dainty; and that in his time the body of a plump girl of fifteen, who was crucified for an attempt to poison the emperor, was sold to his imperial majesty's prime-minister of state, and other great mandarins of the court, in joints from the gibbet at four hundred crowns. Neither indeed can I deny, that if the same use were made of several plump young girls in this town, who without one single groat to their fortunes, cannot stir abroad without a chair, and appear at a playhouse and assemblies in foreign fineries which they never will pay for, the kingdom would not be the worse.

Some persons of a desponding spirit are in great concern about that vast number of poor people, who are aged, diseased, or maimed; and I have been desired to employ my thoughts, what course may be taken to ease the nation of so grievous an incumbrance. But I am not in the least pain upon that matter, because it is very well known, that they are every day dying, and rotting, by cold and famine, and filth and vermin, as fast as can be reasonably expected. And as to the young laborers, they are now in almost as hopeful a condition: they cannot get work, and consequently pine away for want of nourishment, to a degree, that if at any time they are accidentally hired to common labor, they have not strength to perform it; and thus the country and themselves are happily delivered from the evils to come.

I have too long digressed, and therefore shall return to my subject. I think the advantages by the proposal which I have made, are obvious and many, as well as of the highest importance.

For first, as I have already observed, it would greatly lessen the number of papists, with whom we are yearly over-run, being the principal breeders of the nation, as well as our most dangerous enemies; and who stay at home on purpose to deliver the kingdom to the pretender, hoping to take their advantage by the absence of so many good protestants, who have chosen rather to leave their country, than stay at home and pay tithes against their conscience to an episcopal curate.

Secondly, the poorer tenants will have something valuable of their own, which by law may be made liable to distress, and help to pay their landlord's rent; their corn and cattle being already seized, and money a thing unknown.

Thirdly, whereas the maintenance of a hundred thousand children, from two years old and upward, cannot be computed at less than ten shillings a piece per annum, the nation's stock will be thereby increased fifty thousand pounds per annum, beside the profit of a new dish introduced to the tables of all gentlemen of fortune in the kingdom, who have any refinement in taste. And the money will circulate among ourselves, the goods being entirely of our own growth and manufacture.

Fourthly, the constant breeders, beside the gain of eight shillings sterling per annum by the sale of their children, will be rid of the charge of maintaining them after the first year.

Fifthly, this food would likewise bring great custom to taverns; where the vintners will certainly be so prudent as to procure the best receipts for dressing it to perfection, and, consequently, have their houses frequented by all the fine gentlemen, who justly value themselves upon their knowledge in good eating: and a skilful cook, who understands how to oblige his guests, will contrive to make it as expensive as they please.

Sixthly, this would be a great inducement to marriage, which all wise nations have either encouraged by rewards, or enforced by laws and penalties. It would increase the care and tenderness of mothers toward their children, when they were sure of a settlement for life to the poor babes, provided in some sort by the public, to their annual profit or expense. We should see an honest emulation among the married women, which of them could bring the fattest child to the market. Men would become as fond of their wives during the time of their pregnancy as they are now of the mares in foal, their cows in calf, their sows when they are ready to farrow; nor offer to beat or kick them (as is too frequent a practice) for fear of a miscarriage.

Many other advantages might be enumerated. For instance, the addition of some thousand carcasses in our exportation of barrelled beef; the propagation of swine's flesh, and improvement in the art of making good bacon, so much wanted among us by the great destruction of pigs, too frequent at our table; which are no way comparable in taste or magnificence to a well grown, fat, yearling child, which roasted whole will make a considerable figure at a lord mayor's feast, or any other public entertainment. But this, and many others, I omit, being studious of brevity.

Supposing that one thousand families in this city would be constant customers for infants' flesh, beside others who might have it at merry-meetings, particularly at weddings and christenings, I compute that Dublin would take off annually about twenty thousand carcasses; and the rest of the kingdom (where probably they will be sold somewhat cheaper) the remaining eighty thousand.

I can think of no one objection, that will possibly be raised against this proposal, unless it should be urged, that the number of people will be thereby much lessened in the kingdom. This I freely own, and it was indeed one principal design in offering it to the world. I desire the reader will observe, that I calculate my remedy for this one individual kingdom of Ireland, and for no other that ever was, is, or I think ever can be upon earth. Therefore let no man talk to me of other expedients:

of taxing our absentees at five shillings a pound: of using neither clothes, nor household-furniture, except what is our own growth and manufacture: of utterly rejecting the materials and instruments that promote foreign luxury: of curing the expensiveness of pride, vanity, idleness, and gaming in our women: of introducing a vein of parsimony, prudence and temperance; of learning to love our country, in the want of which we differ even from LAPLANDERS, and the inhabitants of TOPINAMBOO: of quitting our animosities and factions, nor acting any longer like the Jews, who were murdering one another at the very moment their city was taken: of being a little cautious not to sell our country and conscience for nothing: of teaching landlords to have at least one degree of mercy toward their tenants: lastly, of putting a spirit of honesty, industry, and skill into our shopkeepers; who, if a resolution could now be taken to buy only our negative goods, would immediately unite to cheat and exact upon us in the price, the measure, and the goodness, nor could ever yet be brought to make one fair proposal of just dealing, though often and earnestly invited to it.

Therefore I repeat, let no man talk to me of these and the like expedients, till he has at least some glimpse of hope, that there will ever be some hearty and sincere attempt to put them in practice.

But, as to myself, having been wearied out for many years with offering vain, idle, visionary thoughts, and at length utterly despairing of success, I fortunately fell upon this proposal; which, as it is wholly new, so it has something solid and real, of no expence and little trouble, full in our own power, and whereby we can incur no danger in disobliging ENGLAND. For this kind of commodity will not bear exportation, the flesh being of too tender a consistence to admit a long continuance in salt, although perhaps I could name a country, which would be glad to eat up our whole nation without it.

After all, I am not so violently bent upon my own opinion as to reject any offer proposed by wise men, which shall be found equally innocent, cheap, easy, and effectual. But before something of that kind shall be advanced in contradiction to my scheme, and offering a better, I desire the author or authors will be pleased maturely to consider two points. First, as things now stand, how they will be able to find food and raiment for a hundred thousand useless mouths and backs. And secondly, there being a round million of creatures in human figure throughout this kingdom, whose whole subsistence put into a common stock would leave them in debt two millions of pounds sterling, adding those who are beggars by profession, to the bulk of farmers, cottagers, and laborers, with the wives and children, who are beggars in effect; I desire those politicians who dislike my overture, and may perhaps be so bold as to attempt an answer, that they will first ask the parents of these mortals, whether they would not at this day think it a great happiness to have been sold for food at a year old, in the manner I prescribe, and thereby have avoided such a perpetual scene of misfortunes, as they have since gone through, by the oppression of landlords, the impossibility of paying rent without money or trade, the want of common sustenance, with neither house nor clothes to cover them from the inclemencies of the weather, and the most inevitable prospect of entailing the like, or greater miseries, upon their breed for ever.

I profess, in the sincerity of my heart, that I have not the least personal interest in endeavoring to promote this necessary work, having no other motive than the public good of my country, by advancing our trade, providing for infants, relieving the poor, and giving some pleasure to the rich. I have no children by which I can propose to get a single penny; the youngest being nine years old, and my wife past child-bearing.

On Population

Thomas Robert Malthus

In reading Mr. Godwin's ingenious work on political justice,* it is impossible not to be struck with the spirit and energy of his style, the force and precision of some of his reasonings, the ardent tone of his thoughts, and particularly with that impressive earnestness of manner which gives an air of truth to the whole. At the same time it must be confessed that he has not proceeded in his inquiries with the caution that sound philosophy requires; his conclusions are often unwarranted by his premises; he fails sometimes in removing objections which he himself brings forward; he relies too much on general and abstract propositions, which will not admit of application; and his conjectures certainly far outstrip the modesty of nature.

The system of equality, which Mr. Godwin proposes, is, on a first view of it, the most beautiful and engaging of any that has yet appeared. A melioration of society to be produced merely by reason and conviction gives more promise of permanence than any change effected and maintained by force. The unlimited exercise of private judgment is a doctrine grand and captivating, and has a vast superiority over those systems, where every individual is in a manner the slave of the public.

From *An Essay on the Principle of Population; or A View of Its Past and Present Effects on Human Happiness*, by Thomas Robert Malthus (London, 1817), pp. 248-270. Reproduced here under the editor's title. Spelling and punctuation have been modernized.

***Enquiry Concerning Political Justice* (1793) by William Godwin was one of the most extreme assertions of rationalistic philosophical anarchism ever made. Godwin condemned the economic organization of society with its extremes of wealth and poverty and believed that universal benevolence would solve all problems. He envisaged a future of human progress based on the assumption that man and society are perfectible.

The substitution of benevolence, as the masterspring and moving principle of society, instead of self-love, appears at first sight to be a consummation devoutly to be wished. In short, it is impossible to contemplate the whole of this fair picture, without emotions of delight and admiration, accompanied with an ardent longing for the period of its accomplishment. But alas! that moment can never arrive. The whole is little better than a dream—a phantom of the imagination. These "gorgeous palaces" of happiness and immortality, these "solemn temples" of truth and virtue, will dissolve, "like the baseless fabric of a vision," when we awaken to real life, and contemplate the genuine situation of man on earth.

Mr. Godwin, at the conclusion of the third chapter of his eighth book, speaking of population, says, "There is a principle in human society, by which population is perpetually kept down to the level of the means of subsistence. Thus among the wandering tribes of America and Asia we never find, through the lapse of ages, that population has so increased, as to render necessary the cultivation of the earth." This principle, which Mr. Godwin thus mentions as some mysterious and occult cause, and which he does not attempt to investigate, has appeared to be the law of necessity—misery, and the fear of misery.

The great error under which Mr. Godwin labors throughout his whole work is the attributing of almost all the vices and misery that prevail in civil society to human institutions. Political regulations and the established administration of property are, with him, the fruitful sources of all evil, the hotbeds of all the crimes that degrade mankind. Were this really a true state of the case, it would not seem an absolutely hopeless task, to remove evil completely from the world; and reason seems to be the proper and adequate instrument for effecting so great a purpose. But the truth is, that though human institutions appear to be, and indeed often are, the obvious and obtrusive causes of much mischief to society, they are, in reality, light and superficial, in comparison with those deeper-seated causes of evil, which result from the laws of nature and the passions of mankind.

In a chapter on the benefits attendant upon a system of equality, Mr. Godwin says, "The spirit of oppression, the spirit of servility, and the spirit of fraud, these are the immediate growth of the established administration of property. They are alike hostile to intellectual improvement. The other vices of envy, malice and revenge, are their inseparable companions. In a state of society where men lived in the midst of plenty, and where all shared alike the bounties of nature, these sentiments would inevitably expire. The narrow principle of selfishness would vanish. No man being obliged to guard his little store, or provide with anxiety and pain for his restless wants, each would lose his individual existence in the thought of the general good. No man would be an enemy to his neighbours, for they would have no subject of contention; and of consequence philanthropy would resume the empire which reason assigns her. Mind would be delivered from her perpetual anxiety about corporal support; and be free to expatiate in the field of thought which is congenial to her. Each would assist the inquiries of all."

This would indeed be a happy state. But that it is merely an imaginary picture with scarcely a feature near the truth, the reader, I am afraid, is already too well convinced.

Man cannot live in the midst of plenty. All cannot share alike the bounties of nature. Were there no established administration of property, every man would be obliged to guard with force his little store. Selfishness would be triumphant. The subjects of contention would be perpetual. Every individual would be under a constant anxiety about corporal support, and not a single intellect would be left free to expatiate in the field of thought.

How little Mr. Godwin has turned his attention to the real state of human society will sufficiently appear from the manner in which he endeavors to remove the difficulty of a superabundant population. He says, "The obvious answer to this objection is, that to reason thus is to foresee difficulties at a great distance. Three-fourths of the habitable globe are now uncultivated. The parts already cultivated are capable of immeasurable improvement. Myriads of centuries of still increasing population may pass away, and the earth be still found sufficient for the subsistence of its inhabitants."

I have already pointed out the error of supposing that no distress or difficulty would arise from a redundant population, before the earth absolutely refused to produce any more. But let us imagine for a moment Mr. Godwin's system of equality realized in its utmost extent, and see how soon this difficulty might be expected to press, under so perfect a form of society. A theory that will not admit of application cannot possibly be just.

Let us suppose all the causes of vice and misery in this island removed. War and contention cease. Unwholesome trades and manufactories do not exist. Crowds no longer collect together in great and pestilent cities for purposes of court ingrigue, of commerce, and of vicious gratification. Simple, healthy and rational amusements take place of drinking, gaming and debauchery. There are no towns sufficiently large to have any prejudicial effects on the human constitution. The greater part of the happy inhabitants of this terrestrial Paradise live in hamlets and farm-houses scattered over the face of the country. All men are equal. The labors of luxury are at an end; and the necessary labors of agriculture are shared amicably among all. The number of persons and the produce of the island we suppose to be the same as at present. The spirit of benevolence, guided by impartial justice, will divide this produce among all the members of society according to their wants. Though it would be impossible that they should all have animal food every day, yet vegetable food, with meat occasionally, would satisfy the desires of a frugal people, and would be sufficient to preserve them in health, strength and spirits.

Mr. Godwin considers marriage as a fraud and a monopoly. Let us suppose the commerce of the sexes established upon principles of the most perfect freedom. Mr. Godwin does not think himself that this freedom would lead to a promiscuous intercourse; and in this I perfectly agree with him. The love of variety is a vicious, corrupt and unnatural taste, and could not prevail in any great degree in a simple and virtuous state of society. Each man would probably select for himself a partner, to whom he would adhere, as long as that adherence continued to be the choice of both parties. It would be of little consequence, according to Mr. Godwin, how many children a woman had, or to whom they belonged. Provisions and assistance would spontaneously flow from the quarter in which they abounded to the quarter in which they were deficient. And every man, according to his capacity, would be ready to furnish instruction to the rising generation.

I cannot conceive a form of society so favorable upon the whole to population. The irremediableness of marriage, as it is at present constituted, undoubtedly deters many from entering into this state. An unshackled intercourse on the contrary would be a most powerful incitement to early attachments; and as we are supposing no anxiety about the future support of children to exist, I do not conceive that there would be one woman in a hundred, of twenty-three years of age, without a family.

With these extraordinary encouragements to population, and every cause of depopulation, as we have supposed, removed, the numbers would necessarily increase faster than in any society that has ever yet been known. I have before mentioned that the inhabitants of the back settlements of America appear to double their numbers in fifteen years. England is certainly a more healthy country than the back settlements of America; and as we have supposed every house in the island to be airy and wholesome, and the encouragements to have a family greater even than in America, no probable reason can be assigned, why the population should not double itself in less, if possible, than fifteen years. But to be quite sure, that we do not go beyond the truth, we will only suppose the period of doubling to be twenty-five years; a ratio of increase, which is slower than is known to have taken place throughout all the United States of America.

There can be little doubt, that the equalization of property which we have supposed, added to the circumstance of the labor of the whole community being directed chiefly to agriculture, would tend greatly to augment the produce of the country. But to answer the demands of a population increasing so rapidly, Mr. Godwin's calculation of half an hour a day would certainly not be sufficient. It is probable that the half of every man's time must be employed for this purpose. Yet with such or much greater exertions, a person who is acquainted with the nature of the soil in this country, and who reflects on the fertility of the lands already in cultivation, and the barrenness of those that are not cultivated, will be very much disposed to doubt, whether the whole average produce could possibly be doubled in twenty-five years from the present period. The only chance of success would be from the ploughing up most of the grazing countries, and putting an end almost entirely to animal food. Yet this scheme would probably defeat itself. The soil of England will not produce much without dressing; and cattle seem to be necessary to make that species of manure, which best suits the land.

Difficult however as it might be to double the average produce of the island in twenty-five years, let us suppose it effected. At the expiration of the first period therefore, the food, though almost entirely vegetable, would be sufficient to support in health the doubled population of 22 millions.

During the next period, where will the food be found to satisfy the importunate demands of the increasing numbers? Where is the fresh land to turn up? Where is the dressing necessary to improve that which is already in cultivation? There is no person with the smallest knowledge of land but would say that it was impossible that the average produce of the country could be increased during the second twenty-five years by a quantity equal to what it at present yields. Yet we will suppose this increase, however improbable, to take place. The exuberant strength of the argument allows of almost any concession. Even with this concession, however, there would be 11 millions at the expiration of the second term unprovided for. A quantity equal to the frugal support of 33 millions would be to be divided among 44 millions.

Alas! what becomes of the picture, where men lived in the midst of plenty, where no man was obliged to provide with anxiety and pain for his restless wants; where the narrow principle of selfishness did not exist; where the mind was delivered from her perpetual anxiety about corporal support, and free to expatiate in the field of thought which is congenial to her? This beautiful fabric of the imagination vanishes at the severe touch of truth. The spirit of benevolence, cherished and invigorated by plenty, is repressed by the chilling breath of want. The hateful passions that had vanished reappear. The mighty law of self-preservation expells all the softer and more exalted emotions of the soul. The temptations to evil are too strong for human nature to resist. The corn is plucked up before it is ripe, or secreted in unfair proportions; and the whole black train of vices that belong to falsehood are immediately generated. Provisions no longer flow in for the support of a mother with a large family. The children are sickly from insufficient food. The rosy flush of health gives place to the pallid cheek and hollow eye of misery. Benevolence, yet lingering in a few bosoms, makes some faint expiring struggles, till at length self-love resumes his wonted empire, and lords it triumphant over the world.

No human institutions here existed, to the perverseness of which Mr. Godwin ascribes the original sin of the worst men. No opposition had been produced by them between public and private good. No monopoly had been created of those advantages which reason directs to be left in common. No man had been goaded to the breach of order by unjust laws. Benevolence had established her reign in all hearts. And yet in so short a period as fifty years, violence, oppression, falsehood, misery, every hateful vice and every form of distress, which degrade and sadden the present state of society, seem to have been generated by the most imperious circumstances, by laws inherent in the nature of man, and absolutely independent of all human regulations.

If we be not yet too well convinced of the reality of this melancholy picture, let us but look for a moment into the next period of twenty-five years, and we shall see that according to the natural increase of population 44 millions of human beings would be without the means of support; and at the conclusion of the first century, the population would have had the power of increasing to 176 millions, while the food was only sufficient for 55 millions, leaving 121 millions unprovided for: and yet all this time we are supposing the produce of the earth absolutely unlimited, and the yearly increase greater than the boldest speculator can imagine.

This is undoubtedly a very different view of the difficulty arising from the principle of population from that which Mr. Godwin gives, when he says, "Myriads of centuries of still increasing population may pass away, and the earth be still found sufficient for the subsistence of its inhabitants."

I am sufficiently aware that the redundant millions which I have mentioned could never have existed. It is a perfectly just observation of Mr. Godwin that "there is a principle in human society by which population is perpetually kept down to the level of the means of subsistence." The sole question is, what is this principle? Is it some obscure and occult cause? Is it some mysterious interference of Heaven, which at a certain period strikes the men with impotence, and the women with barrenness? Or is it a cause open to our researches within our view; a cause which has constantly been observed to operate, though with varied force, in every state in which man has been placed? Is it not misery, and the fear of misery, the necessary and inevitable results of the laws of nature in the present stage of man's existence, which human institutions, so far from aggravating, have tended considerably to mitigate, though they can never remove?

It may be curious to observe, in the case that we have been supposing, how some of the principal laws, which at present govern civilized society, would be successively dictated by the most imperious necessity. As man, according to Mr. Godwin, is the creature of the impressions to which he is subject, the goadings of want could not continue long, before some violations of public or private stock would necessarily take place. As these violations increased in number and extent, the more active and comprehensive intellects of the society would soon perceive that, while the population was fast increasing, the yearly produce of the country would shortly begin to diminish. The urgency of the case would suggest the necessity of some immediate measures being taken for the general safety. Some kind of convention would be then called, and the dangerous situation of the country stated in the strongest terms. It would be observed that while they lived in the midst of plenty it was of little consequence who labored the least, or who possessed the least, as every man was perfectly willing and ready to supply the wants of his neighbor. But that the question was no longer, whether one man should give to another that which he did not use himself; but whether he should give to his neighbor the food which was absolutely necessary to his own existence. It would be represented that the number of those who were in want very greatly exceeded the number and means of those who should supply them; that these pressing wants, which from the state of the produce of the country could not all be gratified, had occasioned some flagrant violations of justice; that these violations had already checked the increased of food, and would, if they were not by some means or other prevented, throw the whole community into confusion; that imperious necessity seemed to dictate that a yearly increase of produce should, if possible, be obtained at all events; that in order to effect this first great and indispensable purpose, it would be advisable to make a more complete division of land, and to secure every man's property against violation by the most powerful sanctions.

It might be urged perhaps by some objectors, that as the fertility of the land increased, and various accidents occurred, the shares of some men might be much more than sufficient for their support; and that when the reign of self-love was once established, they would not distribute their surplus produce without some compensation in return. It would be observed in answer that this was an inconvenience greatly to be lamented; but that it was an evil which would bear no comparison to the black train of distresses inevitably occasioned by the insecurity of property; that the quantity of food, which one man could consume, was necessarily limited by the narrow capacity of the human stomach; that it was certainly not probable that he should throw away the rest; and if he exchanged his surplus produce for the labor of others, this would be better than that these others should absolutely starve.

It seems highly probable therefore that an administration of property, not very different from that which prevails in civilized states at present, would be established as the best (though inadequate) remedy for the evils which were pressing on the society.

* * * *

And thus it appears that a society constituted according to the most beautiful form that imagination can conceive, with benevolence for its moving principle instead of self-love, and with every evil disposition in all its members corrected by reason, not force, would from the inevitable laws of nature, and not from any fault in human institutions, degenerate in a very short period into a society constructed upon a plan not essentially different from that which prevails in every known state at present; a society, divided into a class of proprietors and a class of laborers, and with self-love for the mainspring of the great machine.

In the supposition which I have made, I have undoubtedly taken the increase of population smaller, and the increase of produce greater, than they really would be. No reason can be assigned why, under the circumstances supposed, population should not increase faster than in any known instance. If then we were to take the period of doubling at fifteen years instead of twenty-five years, and reflect upon the labor necessary to double the produce in so short a time, even if we allow it possible; we may venture to pronounce with certainty, that, if Mr. Godwin's system of society were established in its utmost perfection, instead of myriads of centuries, not thirty years could elapse before its utter destruction from the simple principle of population.

I have taken no notice of emigration in this place, for obvious reasons. If such societies were instituted in other parts of Europe, these countries would be under the same difficulties with regard to population, and could admit no fresh members into their bosoms. If this beautiful society were confined to our island, it must have degenerated strangely from its original purity, and administer but a very small portion of the happiness it proposed, before any of its members would voluntarily consent to leave it, and live under such governments as at present exist in Europe, or submit to the extreme hardships of first settlers in new regions.

Toward a Non-Malthusian Population Policy

Jean Mayer

One theme of this essay is that food is only one of the elements in the population problem. Admittedly, at present, it is a major factor in some parts of the world; but there are large areas where the national food supply is a minor factor and others where it is not a factor at all. Furthermore, considering the world as a whole, there is no evidence that the food situation is worsening and there is at least a likelihood that food may at some time (20 or 30 years from now) be removed altogether as a limiting factor to population. Yet, to deny that the population problem is basically one of food for survival is not to deny that there is a population problem; it is in fact to remove the appearance of a safety valve and also to reveal the problem in its generality. For were we really to starve when the population reaches a certain magic number, this in turn would cause a drastic increase in child and infant mortality, decreased fertility, and a shortening of the average life span. In other words, it would cause the increase in population to be self-limiting. If the world can continue to feed—however badly—an ever-increasing number of people, this safety valve (however unpalatable, it *would* be a safety valve) is missing. And if lack of food is not a component of the definition of overpopulation, rich countries as well as poor ones become candidates for overpopulation—now.

Another theme is that there is a strong case to be made for a stringent population policy on exactly the reverse of the basis Malthus expounded. Malthus was concerned with the steadily more widespread poverty that indefinite population growth would inevitably create. I

From *Columbia Forum*, XII:5-13 (Summer 1969).

am concerned about the areas of the globe where people are rapidly becoming richer. For rich people occupy much more space, consume more of each natural resource, disturb the ecology more, and create more land, air, water, chemical, thermal, and radioactive pollution than poor people. So it can be argued that from many viewpoints it is even more urgent to control the numbers of the rich than it is to control the numbers of the poor.

The population problem is not new, although it has recently acquired new and dangerous dimensions. In all early treatments of the subject, considerations of population policy were not closely linked to economic concerns or the availability of food. Plato, who undertook nothing less than the projection of an ideal city-state in estimating the numbers needed for the various functions of citizenship, arrived at the figure of 5,040 citizens as the desirable size, adequate to "furnish numbers for war and peace, for all contracts and dealings, including taxes and divisions of the land." In *The Republic* he described his well-known eugenics proposal for public hymeneals of licensed breeders. His preoccupation was with the quality of man and of the state; not with the availability of food and other resources. Aristotle, who was concerned with certain of the economic consequences of overpopulation, though not specifically with food, warned in *Politics* that "a neglect of an effective birth control policy is a never-failing source of poverty, which is in turn the parent of revolution and crime," and advised couples with an excessive number of children to abort succeeding pregnancies "before sense and life have begun."

Plato and Aristotle did not go unchallenged. The Pythagoreans, in particular Hippocrates, opposed abortion. The Hippocratic oath contains the pledge: "I will not give a woman an abortive remedy." Of greater subsequent importance, the Romans, particularly Cicero, disapproved of the Greek views on population. They were not so much concerned with the quality of man as with the excellence of the Empire. Rome taxed celibacy and rewarded large families. Roman ideas, incidentally, were very similar to those of Confucius and his followers, also citizens of a large and expanding empire, and equally convinced that a numerous and expanding population should be promoted by wise rulers. The economic consequences of large populations were essentially ignored by the Romans; Confucius dealt with them by enunciating the rather intriguing formula: "Let the producers be many and the consumers few."

The Hebrews and the Fathers of the Church were similarly uninterested in the economic implications of population growth. Biblical and early Christian writers can, indeed, hardly be considered to have had a population policy, though their concepts of family life and of the dignity of man are as basic now as they were milleniums ago. Children were repeatedly designated as the gifts of God, with large families particularly blessed. The prescriptions of Saint Paul were somewhat more complex: while he stated that women could merit eternal salvation through bearing children if they continued to be faithful, holy, and modest, he praised virginity as more blessed than marriage, and dedicated widowhood as preferred above remarriage. Following Saint Paul, the position of Christians in sexual relations became variegated: from strict antinomianism, believers in all possible experiences, to fanatical ascetics who believed in self-castration to remove all possibility of temptation, with the center of gravity of Christian opinion somewhere between the traditional Roman double standard (strict virtue expected from the woman, somewhat more permissive rules for the man) and the more puritanical ideal of the Stoics.

In this ethical Babel, the absence of strict theological canons made it possible for intelligent citizens of a crowded and decaying empire to discuss the possible economic and political consequences of overpopulation. In *De Anima* Tertullian wrote: "The scourges of pestilence, famine, wars, and earthquakes have come to be regarded as a blessing to crowded nations, since they served to prune away the luxuriant growth of the human race." The position of the Church against abortion hardened in the third century, when Saint Hippolytus opposed Pope Calixtus I for showing too much leniency toward the abortionists, and reiterated the Christian position that the fetus is a person and not, as in Roman Law, a part of the mother. Saint Augustine, rising from a Manichaean background and a personally unhappy sexual history, defined the purpose of Christian marriage as procreation, with abstinence permissible by mutual consent. This basis of the Christian marriage, unmodified by Thomas Aquinas or the medieval theologians, unmodified by Luther (an Augustinian, very much attached to the pattern of the order) was to survive far into the twentieth century.

In spite of theologians, Tertullian was echoed 1,300 years later in Botero, a sixteenth-century Italian writer, who held that man's productive powers are inferior to his reproductive powers, which do not diminish automatically when population increases. The population of the world, then, must be constantly checked by war and epidemics, the earth already holding as many people as it can feed. From Botero onward, concepts of optimum population size became indissolubly linked to economic considerations, but to economic considerations of the lowest order. Population limitation was advocated by writers, Malthus foremost among them, who felt they could demonstrate that population will inevitably rise to the very margin of food production capacity, with misery and vice the only consequences. The examples chosen were often unfortunate in the light of hindsight: Malthus based his prediction on an examination of the United States of the late eighteenth century. On the other side, mercantilist writers and rulers once again saw an increase in population as a guarantee of ample manpower for production and for war, and as a test of good government. Through the nineteenth century the debate continued. Malthusians saw the solution of economic problems due to overpopulation in continence and in more poverty—specifically, the repeal of the poor laws. The belief in the inevitability of starvation and the desirability of a *laissez faire* policy was in no small measure responsible for governmental inaction during the Irish famine. At the other end of the spectrum, Marx and Engels opposed Malthus as a peculiarly vicious and obsolete defender of capitalism. "Overpopulation" was a bourgeois invention designed to justify the poverty of the working classes. Improved production and distribution, not restriction of births, was the answer. A socialistic economy could thrive under all conditions of population growth, while an economy based on scarcity and high prices required birth control to mitigate its glaring deficiencies. Oddly enough, a number of modern Catholic philosophers have held a viewpoint not very different from that of Karl Marx.

The position of many Chinese leaders at present is a combination of orthodox Marxist anti-Malthusianism and traditional Chinese predilection for large families. Others, echoing Francis Place and the liberal socialists, advocate the availability of the means of birth control so as to permit the liberation of women and their participation in the edification of socialism (being careful meanwhile to avoid any Malthusian implication).

Since the mid-nineteenth century, three profound revolutions have been taking place: a technological revolution, which promises to accelerate food production still faster; a demographic explosion, which is also accelerating and places the problem of population in an even more dramatic context; and changes in human attitudes, for which Harlan Cleveland has coined the felicitous expression, "the revolution of rising expectations." It is the contention of this writer that nothing is more dangerous for the cause of formulating a sound population policy than to approach the problem in nineteenth-century terms. By continuing to link the need for population control to the likelihood that food supply will be increasingly limited, the elaboration of birth control programs of sufficient magnitude will be held up for many years, perhaps many generations. In contemporary terms, it may well be that the controversy between Plato and Cicero makes more sense than that between the neo-Malthusians and the neo-Marxists.

That the magnitude of the population problem has increased dramatically in recent years is well publicized. Scholars have estimated that after hundreds of thousands of years of slow growth, the population of the world reached the quarter billion mark some time around the beginning of this era. It doubled to 500 million by 1650. Two centuries later it reached the billion mark. The next doubling took 80 years, with a population of 2 billion in 1930. It would appear that the world is on its way to the next doubling, to 4 billion in 45 years, by 1975; and a population of 8 billion may well be reached within the following 30 or 35 years unless rates of growth are drastically decreased. The present growth rate would lead to a population of 500 billion by the year 2200, and give the surface of all continents a population density equal to that of Washington, D.C., at present!

This increase has been due not to an increase in birth rates, but to a decrease in death rates. Around 1700, life expectancy at birth of European populations was about 33 years, and had increased little in the previous three to four hundred years. By 1950, life expectancy in Western and Central Europe and in the United States had increased to 66-69 years, an increase of over 100 per cent. This decrease in mortality rates is no longer confined to populations of European stocks. In 1946, the death rate of the Moslem population of Algeria was higher than that of Sweden in 1775. In 1954, in spite of generalized guerilla war on its territory, the death rate of this population was lower than that of Sweden in 1875. A similar telescoping of the drop in death rates is going on all over the world.

From a demographic point of view it must be noted that a drop in the death rate, with birth rate unchanged, not only results in an increase in the rate of population growth, but also produces an acceleration in the rate of growth itself: a decline in age-specific mortality rates in ages prior to the end of the childbearing age has the same demographic effect as an increase in the birth rate. In the United States, 97 out of every 100 newborn white females reach the age of 20; 91 reach the age of 50. In Guatemala, only 70 reach the age of 20; 49 that of 50. If the death rate in Guatemala fell within the next decade to somewhere near the 1950 United States level, a not unlikely development, this alone would increase the number of women reaching the beginning of the childbearing period by 85 per cent. Because of the high proportion of young people in underdeveloped countries generally—a country like Costa Rica has twice the proportion of people under 15 that Sweden has—this drop in the death rate in the pre-childbearing period has now and will have in the next few years a gigantic effect on the birth rate. Brazil had 52 million people in 1950, 71 million in 1960, and 83 million in 1966. If present rates prevail it should have 240 million by the year 2000, or

14 times the 1900 population. With a drop in mortality in the young age groups, the increase could be even more spectacular.

The significance of the demographic trends within this country is not generally appreciated. The United States, with a population of 200 million, has at present one-sixteenth of the earth's population on one-sixteenth of the land area. Though a number of underdeveloped areas are piling up population faster, we are accumulating about 2.2 million people per year, more than any increase before 1946. The rate of growth seems unimpressive, 1.1 for the year 1967 (the highest rate reached was 1.8 in 1946 to 1957). If the rate prevailing over the past five years persists, the population of the United States will reach 300 million by the year 1990. What most of us have tended to ignore is that the so-called baby boom of the postwar era followed a period of depression and very low birth rates: from 1920 to 1933 the birth rate had fallen steadily from 27.7 per 1,000 in 1920 to 18.4 in 1933. The absolute decline in births was less steep, because the numerical base of women of childbearing age was still growing. When the birth rate started rising in the early forties, the increase was applied to the still large number of women born between 1916 and 1924. Since 1945, the baby boom that has been so well publicized had actually been taking place on the basis of the shrinking group of women of childbearing age born since 1924. As of 1963, the last of the undersize groups had entered the reproducing age. From 1964 (when the first girls born in the big postwar years reached the age of 18), the number of women in the childbearing age has started increasing rapidly. While in 1940 there were 32 million women 15 to 44 years of age, in 1950 34 million, and in 1960 36 million (a very slow increase), there will be 43 million in 1970 and 54 million in 1980. While the birth rate is declining (and while a better index, the age-standardized general fertility rate based upon women of childbearing age only is also declining), the sheer existence of the number of women and girls alive now means that even in the unlikely event that the fertility rate fell to the historical lows of the depression years and never departed from it, the population of the United States would still more than double in the next century. The reader will, I trust, give me credit for not minimizing the problem of total population either at home or for the world at large.

With this picture of ever-increasing numbers of people, the first reaction among a portion of the public is that we are running out of space, that the "population density" is becoming dangerously high. This concept of "population density"—number of people per unit surface—has underlain the concept of "overpopulation" in the past. It is not very useful except where the primary resources are extractive (mining) and where the most primitive types of agriculture (independent of industry for fertilizers, machines, etc., and hence essentially dependent on area) and forestry prevail. It also pre-supposes that there is no industry to absorb surplus manpower. It is a concept of dubious value where non-extractive industries are dominant and where trade is possible. The high density band from Boston to Washington has an area of 14,000 square miles, an aggregate population of over 30 million (or over 2,000 persons per square mile), and very limited natural resources. The median family income is $1,000 more than for the United States as a whole. Can this area be said to be overpopulated from a material standpoint? To those who object that this area is part of a larger and less densely populated whole, one might point to prosperous Holland, or Belgium, or even Hong Kong, which, although trade with its hinterland is very meagre (imports from mainland China represent only 17 per cent of total imports), not only houses 3.1 million people on 398 square miles (12,700 per square mile), but has shown an unexcelled increase in national product of 7 to 10 per cent per year—a doubling of real output within 10 years. Once one argues that a certain population density should be preserved, such as density with respect to capital for example, one is dealing with a much more complex concept. From it follows the idea that some sparsely settled countries need rapid increases in population, preferably through immigration, for optimal use of resources. The mental image of population density entertained by most people is, in any case, complicated by esthetic and social considerations, and "high density" is more likely to be ascribed to Calcutta than to Paris, to Costa Rica than to Denmark.

This leads us to the second and more popular concept, that overpopulation can best be appraised with respect to food resources and that the present rate of increase in the world's population is rapidly carrying us to the brink of or to actual starvation. It is my contention that this is not happening. Furthermore, I do not consider that my belief, which I shall now endeavor to justify, makes me an "optimist" as compared to the legions of conservationists, social scientists, etc., who have embraced a Malthusian "pessimism." If anything, this view makes me even more pessimistic about our chances of limiting the world's population at an early date: famine or the threat of famine is perhaps the worst method of limitation, but it would work.

World War II was not a Malthusian check. In spite of the horrendous numbers of soldiers and civilians killed, in spite of the massive genocide perpetrated by the Nazis, food production decreased much more than population. By 1945, intake per capita was 16 per cent lower than the 1934-38 average. The creation of the Food and Agriculture

Organization, a specialized United Nations Agency that was endowed during its first years with particularly articulate spokesmen, dramatized the worldwide concern over the food situation. The difficulties inherent in getting agriculture going while industry and the means of communication were not yet rebuilt, led to a generalized feeling of pessimism. Cereals, oils, meat, dairy herds were, in succession, the objects of great attention, the conclusion being in each case that pre-war levels of production and consumption were not going to be reached for years. The chaotic state of international trade accentuated shortages, which UNRRA and various emergency agreements attempted to cope with on an *ad hoc* basis. And yet very quickly the situation improved. The oil shortage vanished first; while the gigantic ground nut scheme of the British government, which was supposed to mitigate it, was taking off to a very slow start, the reappearance in the channels of trade of adequate amounts of fats and oils eliminated the motivation for the scheme itself. United States production of cereals and animal products, which had grown during the war in spite of the lack of abundant manpower and the diversion of the chemical industry to military purposes, had to be slowed down as surpluses started accumulating, and, with their appearance, the threat of a collapse of agricultural prices loomed. By 1952-53, the worldwide rate of per capita production of food had overtaken pre-war rates. Since then, the average rate of increase in the production of food for the world at large has been 3 per cent per year, while the population has increased on the average 1.7 per cent. In document No. 8148, the Department of State estimates that if individual consumption levels remained at the 1955-57 level, the world at large would show by 1975 an annual surplus of 40 million tons of wheat and 70 million tons of rice. (This estimate is based on the postulate that there will be no increase in rice production in Europe and North America, and no increase in wheat production in North America.) Actually, this slight but steady gain of food production over population is part of a secular trend. E. S. and W. S. Woytinski, in their monumental *World Population and Production*, estimate that since 1850 the increase in output has been more rapid than the increase in population.

As chairman of the National Council on Hunger and Malnutrition in the United States I have been talking of these evils at home for years. I have done extensive work in malnutrition in Asia and in Africa and have just returned from a trip to Nigeria and to Biafra, where I went to study the famine and the means to alleviate it. I am, therefore, as well aware of the widespread character of malnutrition as anyone in the world. Caloric undernutrition is still found in many parts of the world, and not always as a result of war or civil disorder, earthquakes or floods, invasions of insects and other parasites, or abnormally prolonged droughts. Protein deficiency—kwashiorkor where it occurs without accompanying caloric deprivation; marasmus when both caloric and protein intakes are inadequate—is encountered in varying degrees of prevalence among the young children of most countries of Asia and Africa and in many of Central and South America. Vitamin A deficiency is perhaps underestimated as a threat to the life, and the sight, of children of most of the same areas where protein deficiency is also seen. Riboflavin deficiency, thiamine deficiency (beri-beri in its various forms), and a number of other deficiencies are still very much with us. Still, there is no evidence that the situation is getting worse. The food balance sheets on which postwar pessimism was based are imperfect instruments. As an officer of FAO, I spent considerable time attempting to gauge such unknowns as figures for waste at the retail level and within families, and that portion of the food supply that does not move within the channels of trade (food grown by the farmer for his family is very inaccurately known, particularly as regards fruits and vegetables which tend to be underestimated). The nutritional standards against which available supplies are gauged are themselves being refined. As the results of additional experimental and clinical work become available, it is realized that a number of such standards—those for protein and calcium among others—were probably unnecessarily high. Even without such reevaluation, the evolution of food balance sheets, the only instruments we have to judge the race between food and population, make it apparent that most regions do show the same slow increase of per capita supplies exhibited by the world at large. It must be recognized, of course, that many of the worst nutritional scourges of mankind have been historically due as much to ignorance and to callousness as to lack of nutrients as such. Thousands of children die of protein deficiency in areas where the proteins which would save them do in fact exist and are often consumed in sufficient amounts in the very households where infants and toddlers die for lack of them. A faulty understanding of a child's needs may be the main reason he is denied some of the food consumed by his father and older siblings. As for man's inhumanity to man and its contribution to starvation, it could be illustrated by thousands of examples: cereals being shipped from Ireland under the protection of naval guns during the famine; stocks being withheld during the Congo famine to keep prices up; crop destruction policies in South Vietnam; the food blockade of Biafra.

Certainly as far as food is concerned ours is not one world. The United States government rents 20 million acres from our farmers so that they will not grow food on them. A study made at Iowa State University a few years ago suggests that sixty-two and a half million acres ought to be similarly retired so that surpluses will not continue to be created in relation to the present market. Australia, Canada, New

Zealand, Argentina, and France have been, or are at present, involved in similar efforts to restrict production.

Nor is this idling of food production restricted to highly developed countries. A recent study estimates that Ghanaian farmers work only an average of two hours a day in the cocoa area, the wealthiest agricultural area of the country.

It is fair to say that in most areas of the world the race between food and population would be more favorable to the development of adequate nutrition if the rate of population growth was decreased. But I believe that there are no grounds for saying in 1969 that the nutritional state of the world is getting worse. It is not. And I believe that improvement in communication, availability of surpluses in certain countries, the existence of solid international organizations, and the gradual improvement in international morality make large-scale famines, such as the Irish or the Bengali famine, less likely to occur in this era—except perhaps in Red China because of its alienation from the two richest blocs of countries. (It appears, moreover, that the food situation in China has improved considerably in the past two years, making the recurrence of famine there, as in India, more remote.)

Bad as it is, the present is no worse than the past and probably somewhat better. But what of the future? In absolute numbers, the increase in population is likely to accelerate for some time. Can the food supply be kept up? My contention is that for better or for worse it can and will.

First, let us consider conventional agriculture. FAO's figure indicate that 3.4 billion acres are at present under cultivation. This represents less than 11 per cent of the total land area of the world. Some experts—Prasolov, Shantz, Zimmermann—estimate the area that can eventually be made arable at from 13 to 17 billion acres. Colin Clark, director of the Agricultural Economics Research Institute of Oxford, uses the figure of 19 billion acres, but counts double-cropped tropical lands twice. (He considers, incidentally, that if the land were farmed as well as the Dutch farmers work their acres today, it would support 28 billion people on a Dutch diet; if Japanese standards of farming and nutrition were used, this area would support 95 billion people.)

The biggest potential increase of food production does not, however, come from the extension of the area under cultivation, but from the increase in the use of fertilizers. The phenomenal increase in food production in this country has actually been performed with a reduction in acreage farmed. By pre-World War I standards of cultivation, it took one-and-one-half acres to support an American. If such standards prevailed today, we would need to add at least 40 million acres to our farm area every ten years, or the equivalent of an additional Iowa every decade. In fact, we use fertilizers instead. One ton of nitrogen is the equivalent of 14 acres of good farmland. The use of between two and three hundred thousand tons of nitrogen (and corresponding amounts of other necessary elements) per decade has obviated the need to discover another Iowa. And our use of fertilizer is less intensive than it is in Japan, where it is well over twice ours, or in Western Europe. (Incidentally, in spite of its already high standards of cultivation, Japan is still increasing its agricultural production at a rate of 3 per cent per year.) India, Africa, and most of Latin America use only an infinitesimal fraction of Japanese or Western amounts of fertilizer or none at all. Garst has estimated that an expenditure of ten dollars an acre per year for fertilizers would alone add 50 to 100 per cent to the low yields in underdeveloped countries. Applying this investment to an area of 1.5 billion acres would be the equivalent to adding at least 750 million acres to the crop areas of these countries, the equivalent of a continent bigger than North America. It is interesting to note that this primacy of fertilizers was recognized relatively late. In this country, the recognition dates back only to World War II, and has accelerated since the Korean conflict. In Japan, it dates back to 1950 or thereabout. And the leaders of the U.S.S.R. only recently realized that a large-scale increase in fertilizer output would be easier and more rewarding than the extension of cultivation to the "virgin lands."

There are many other advances in agriculture that have yet to be applied on a large scale. The identification of necessary trace elements and their incorporation into fertilizers and feeds have opened vast areas to cultivation and husbandry in Australia and elsewhere. Selective breeding of plants and animals has permitted the development of species with superior hardiness and increased yields. In the greater part of the world such work has hardly begun. Advances in animal health and nutrition have permitted the mass production of milk and eggs in indoor conditioners on a scale that was unimaginable a few years ago. The city of Los Angeles, for instance, is now an important and efficient dairy area. In some large installations, computers programmed to calculate the cheapest method of providing a diet of known energy and known content in ten essential amino acids, total protein, and other nutrients, automatically set the controls that will mix basic staples providing the cheapest adequate poultry diet as they are informed of the latest commodity prices. Herbicides increase yields; pesticides prevent losses from rodents, insects, and fungi. In many underdeveloped countries one quarter of the crop is lost before it reaches the consumer. Certain methods of preservation of foods by radiation have just been approved by the Food and Drug Administration. The control of weather by seeding clouds for rain; speeding cloud formation by heating lakes by atomic energy; the desalinization of brackish water by various methods, are entering the realm of practical feasibility.

Powerful though these methods of "classical" agriculture are, I believe that they will, within the lifetime of most present inhabitants of this planet, be left far behind as methods of food production. The general public is still unaware of some new developments, their promise, and the extent of the means likely to be expended in the next decade in bringing the results of research to practical application. Large-scale manufacture of food from petro-chemicals started during World War II, when the Germans manufactured synthetic fats to feed forced labor groups. These fats did not conform to desirable standards of taste or safety (they contained a high proportion of branched-chain fatty acids not normally found in nature and probably not fully metabolized, and retained a petroleum-like odor). After the war, interest in "synthetic" fats persisted for a while during the years when it appeared that a shortage of natural fats was likely to be protracted. During the fifties, little or no work was done in this field, but recently some of the larger international oil companies have again become actively interested, and pilot plants are now in operation. Fatty acids, triglycerides (the constituents of our common oils and fats), and fully metabolizable simpler compounds, such as 1,3-butanediol, may soon be manufactured at very low cost for human food and animal feeds. While the promise of abundant and cheap atomic power, widely heralded for the morrow in the more immediate postwar period, has shown itself slow to be realized, it is coming, and it may well be that oil will be increasingly a raw material for food and plastics rather than a fuel.

As a potential source of food production, photosynthesis can be used much more efficiently in algae than in higher plants. With proper mineral fertilization and with the proper rate of removal of the finished products, one square meter may serve to support algae production sufficient to feed one man. And a large proportion of the calories produced—as much as one half—are derived from protein; vitamins are also produced into the bargain. Several universities are working with a number of species, chlorella in particular, and large industrial firms are yearly becoming more interested. The problems entailed in passing from the theoretically feasible to the economically feasible are formidable, but their solution is likely to be hastened for an unexpected reason. Interplanetary travel of long duration and the organization of distant stations require not only recycling of oxygen and waste water; they necessitate the fabrication of food and its integration into the recycling of oxygen, water, and excreta. Over the next two decades, an increasing fraction of the several billion dollars that the United States and the Soviet Union will spend every year for space travel is going to be channeled into life support systems. The money spent in the aggregate on new methods of food production will probably, during that period, dwarf the cost of the Manhattan Project. In many ways, we may have in space exploration what William James called "the moral equivalent of war." We will probably also have in it the technological equivalent of war, without the corresponding losses in men and in resources. The usable "fall-out" of such research is likely to be enormous. Certainly if economical harnessing of photosynthesis, through biological units or directly, can be realized under the hostile interplanetary, lunar, or martial conditions, it should become relatively easy to put it into effect on earth. All this is no longer science fiction. It is as much of a reality as the federal income tax. Obviously, a breakthrough in this field could for centuries altogether remove food as a limiting factor to population growth.

I hope I have said enough to show how dangerous it may turn out to link the population problem so closely to food, as so many writers have done. These have generally been conservationists and social scientists rather than agricultural or nutritional scientists, concerned—rightly—with the effects of crowding which they had observed. At the same time, not sure that the public and governments would agree with them that there was cause for concern, and action, based on these grounds, they have turned to the threat of a worldwide shortage of food as an easily understood, imperative reason for a large-scale limitation of births. Had they consulted nutritionists, agriculturists, and chemists, they might have chosen a more appropriate battleground. For if we can feed an ever-increasing number of people—even if we feed them as badly as many of our contemporaries are fed—their argument fails. And yet there is a need for the establishment as soon as possible of a sound population policy for the world at large.

There is, of course, another good reason for not tying population control to food: this tie eliminates from contention rich countries, and in particular surplus countries such as ours. Our population is increasing faster than it ever has; our major nutrition problem is overweight, our major agricultural problem is our ever-mounting excess production. Does anyone seriously believe this means that we have no population problem? Our housing problems; our traffic problem; the insufficiency of the number of our hospitals, of community recreation facilities; our pollution problems, are all facets of our population problem. I may add that in this country we compound the population problem by the migratory habits of our people: from rural farm areas to urban areas and especially to "metropolitan" areas (212 such areas now have 84 per cent of our population); from low income areas to high income areas; from the East and Midwest to the South and Southwest; from all areas to the Pacific Coast; from the centers of cities to suburbs, which soon form gigantic conurbations, with circumstances everywhere pushing our Negroes into the deteriorating centers of large cities. All this has occurred without any master plan, and with public services continually lagging behind both growth and migrations.

Let us conclude with one specific example: 4 million students were enrolled in U.S. colleges and graduate schools in 1960; 6 million in 1965. The Bureau of the Census estimates that 8 million will seek admission or continued enrollment in 1970; 10 in 1975; 12 in 1980. No one questions our ability to feed these youngsters. But are we as a nation at all prepared for a near doubling of the size of our colleges and universities in 11 years?

Let us now examine the other argument, that in certain ways the rich countries are more immediately threatened by overpopulation. A corollary of this is that the earth as an economic system has more to fear from the rich than from the poor, even if one forgets for a moment the threat of atomic or chemical warfare.

Consider some data from our own country. We have already said that "crowding" is certainly one of the pictures we have in mind when we think of overpopulation. The increased crowding of our cities and our conurbations has been referred to, but what of the great outdoors? In 1930 the number of visitor-days at our national parks was of the order of 3 million (for a population of 122 million); by 1950 it was 33 million (for a population of 151 million); by 1960 it was 79 million (for a population of 179 million); by 1967, 140 million (for a population of 200 million). State parks tell the same story: a rise in visitor-days from 114 million in 1950 to 179 million in 1960, an increase in attendance of over 125 per cent for a rise in population of less than 20 per cent! Clearly, the increase in disposable income (and hence in means of transportation and in leisure) becomes a much more important factor in crowding and lack of privacy than the rise in population.

Not only does the countryside become more rapidly crowded when its inhabitants are rich, it also becomes rapidly uglier. With increasing income, people stop drinking water as much: as a result we spread 48 billion (rust proof) cans and 26 billion (nondegradable) bottles over our landscape every year. We produce 800 million pounds of trash a day, a great deal of which ends up in our fields, our parks, and our forests. Only one third of the billion pounds of paper we use every year is reclaimed. Nine million cars, trucks, and buses are abandoned every year, and while many of them are used as scrap, a large though undetermined number are left to disintegrate slowly in backyards, in fields and woods, and on the sides of highways. The eight billion pounds of plastics we use every year are nondegradable materials. And many of our states are threatened with an even more pressing shortage of water, not because of an increased consumption of drinking fluid by the increasing population, but because people are getting richer and using more water for air-conditioning, swimming pools, and vastly expanded metal and chemical industries.

That the air is getting crowded much more rapidly than the population is increasing is again an illustration that increase in the disposable income is perhaps more closely related to our own view of "overpopulation" than is the population itself. From 1940 to 1967 the number of miles flown has gone from 264 million to 3,334 billion (and the fuel consumed from 22 to 512 million gallons). The very air waves are crowded: the increase in citizen-licensees from 126 thousand to 848 thousand in the brief 1960-67 interval is again an excellent demonstration of the very secondary role of the population increase in the new overpopulation. I believe that as the disposable income rises throughout the world in general, the population pressure due to riches will become as apparent as that due to poverty.

I trust that I have demonstrated how dangerous it is to link constantly in the mind of the public the concept of overpopulation with that of undernutrition. I believe that it is dangerous to link it necessarily with poverty. It is absurd on the basis of any criterion of history, economics, or esthetics. Some countries are poor and densely populated. A few countries are poor and so sparsely populated that economic development (e.g. road-building, creation of markets) becomes very difficult. It is easy to demonstrate that a couple with many children will be unable to save and invest. It is perhaps also true that, as the comparison to nineteenth-century France, England, and Germany suggests, at a certain stage of development, too low a birth rate (as in France then) decreases the ambition and labor of part of the population so that the savings expected from the decreased birth rate never materialize. (Losing wars because of a smaller population and having to pay a heavy tribute, as happened to the French at the conclusion of the 1870-71 war, also nullified this advantage). The fact is that we are not yet in one world and that while in general it is true that population increases make improvement in nutrition and in delivery of social services more difficult, the relation of changes in wealth to changes in population has to be examined in each area on its own merits.

We have seen, furthermore, that there is more to the problem of population than the decrease in income consequent to overpopulation. We have seen that the increase in disposable income creates a population problem that is becoming every day more acute. The ecology of the earth—its streams, woods, animals—can accommodate itself better to a rising *poor* population than to a rising *rich* population. Indeed, to save the ecology the population will have to decrease as the disposable income increases. If we believe, like Plato and Aristotle, in trying for excellence rather than in rejoicing in numbers, we need a population policy now, for the rich as well as the poor. Excellent human beings will not be produced without abundance of cultural as well as material resources and, I believe, without sufficient space. We are likely to run out of certain metals before we run out of food; of paper before we run out of metals. And we are running out of clear streams, pure air, and the familiar sights of Nature while we still have the so-called "essentials" of life. Shall we continue to base the need for a population policy on a nutritional disaster to occur at some hypothetical date, when it is clear that the problem is here, now, for us as well as for others? Shall we continue to hide the fact that a rational policy may entail in many countries not only a plateauing of the population to permit an increase in disposable income, but a decrease of the population as the disposable income rises?

A Naturalist Looks at Overpopulation

Joseph Wood Krutch

One of the many indications that the population explosion poses the most desperate problem of our day is the fact that it inevitably arises in connection with every approach to the analysis of our civilization and its prospects. To the critic of culture it is a part of our emphasis on quantity rather than quality. To the economist it raises the question of economic stability. To the political scientist it evokes the specter of wars for Lebensraum; to the conservationist the equally terrifying specter of universal starvation.

To the specialists in their various fields I leave the discussions appropriate to them and say only something about the situation as it appears to a naturalist; to one who is, of course, aware of its other aspects but tends to think first of man's place in nature and the consequences of modern man's refusal to accept the fact that he is indeed part of a scheme which he can to some extent modify but which he cannot supersede by a scheme of his own making.

It is true, of course, that man became man rather than simply a member of the animal kingdom when he ceased merely to accept and submit to the conditions of the natural world. But it is also true that for many thousands of years his resistance to the laws of animal nature and his modifications of his environment were so minor that they did not seriously interfere with natural law and required no such elaborate management of compensating adjustments as became necessary as soon as his intentions, desires, and will became effective enough to interfere with the scheme of nature.

From *Our Crowded Planet*, edited by Fairfield Osborn (New York, 1962), pp. 207-213.

It was not until well into the nineteenth century that his interferences did become extensive enough to force a dawning realization of the fact that you cannot "control nature" at one point without taking steps to readjust at another the balance which has been upset. Improved methods of agriculture exhaust the soil unless artificial steps are taken to conserve and renew it. You cannot destroy all the vermin without risking the destruction of useful animals. You cannot, as we are just discovering, poison noxious insects without risking the extinction of birds who are an even more effective control. It is not that we should not interfere with nature, but that we must face the consequences of this interference and counteract or ameliorate them by other interferences. You dare not, to put it as simply as possible, attempt to manage at one point and to let nature take her course at another.

Considered in connection with this fact the population explosion becomes merely a special (and especially ominous) example of a phenomenon characteristic of civilized man's peculiar place in nature where he is the only creature capable of effectively interfering with her operations while he remains at the same time not wise enough always to foresee the unwanted consequences of his interference. To reduce it again to the simplest possible terms, he has interfered with nature by preserving individual lives far more successfully than nature had ever been able to preserve them at the same time he has allowed nature to take her course so far as propagation is concerned. As a consequence either one of two things must happen. Either he must control birth as well as death or nature will step in and by her own rough but effective methods—starvation, disease, or the brutal death struggle for food and living room—eliminate the excess which failure to manage the consequences of his management has produced. No matter what fantastic increases technology may bring in the number of men the earth is able to support, the limit must be reached sooner or later.

Every ecologist knows that nature left to herself works out a balance of populations adjusted to the available space and food supply, and that this balance, which involves the various aspects of competition including the predator and his prey, is often remarkably stable over long periods of time. But every ecologist knows also that it may be disturbed and then destroyed by what might appear to be the very slight intervention of man. Introduce and then forget a few goats into the biota of an isolated island, and in a few years nothing but goats—many of them starving—will remain. Nature is efficient but slow. It takes centuries for her to work out a balance. Man can in a few decades make a desert which nature cannot reclaim in centuries. So it is also with instinct, which is geared to millennia, while consciously directed purpose is effective within a few years. The instinct which tells us that the more children we can produce the better, developed in us when man was dominated by nature. It persists fatally in a world he has come to manage and mismanage.

Early proponents of planned parenthood assumed that once easy and reliable methods of birth control were available, convenient, and legal the only remaining impediment to a rational solution of the problem would be that of religious or moral resistance. But in my opinion the existence of the ancient instinct deep in the biological organism is a more formidable enemy than religious dogma. In the United States at least the population has been increasing at an accelerated rate at the same time that methods of birth control have become better known and more readily accessible. The only possible explanation is simply that people continue to *want* more children than is desirable now that the mortality rate has been so greatly reduced. Many who are intellectually convinced that population growth should be reduced nevertheless rejoice in at least their own large families because the impulse to increase and multiply was an instinct long before it was a biblical injunction. Man the thinker lags behind man the technician. No less important is the fact that his instincts lag, not years but millennia, behind even his thinking. The most crucial question is not can he be made to *believe* that too many children are undesirable, but can he conquer his instincts sufficiently to make him *feel* what his intellect has convinced him of?

So much for the special ways in which the naturalist sees the problem. He tends also to be more acutely aware than others of a particular aspect of the undesirable consequences of overpopulation. Many sociologists and political scientists recognize the fact that the question is not simply how many people the earth could possibly support, but what is the optimum number from the standpoint of the possibility of a good life. Just as it is foolish to ask what is the largest number of children a family could possibly consist of rather than how many constitute an ideal family unit, so it is foolish to ask how many could be crowded onto our globe rather than what number can live happily there. Men need not only food and a place to sleep but also room to move about in. It is at least possible to believe that cities are already too big and that life would become almost intolerable if they were both more densely crowded and so merged one with another that there was no escaping from them.

Of this the naturalist is often more acutely aware than either the sociologist or the political scientist because he is more completely convinced than they sometimes are that the best life for the human being is one which is led, partly at least, in the context of nature rather than in a context which consists exclusively of the man-made environment. For a large part of the existing human race in the centers of civilization, contact with the natural world is tending to diminish almost to the vanishing point while he has little experience with anything except bricks, steel and concrete on the one hand and mechanical contrivances on the other. As the cities spread and the country shrinks he is more and more imprisoned with his fellows in a world that has ceased to be even aware of many of the things of which he was once an intimate part. Already he has pushed into extinction many of the creatures with which he once shared the earth.

Those who feel that he has already begun to suffer from this fact, talk about recreational areas, about nature education, about national parks and even about wilderness areas. To some extent they can still meet the objections of those who say that we cannot afford to forego the use of any of our forests, or mountains, or deserts. But if our population continues to grow at its present rate, it will soon become evident that we do indeed need every available acre of possibly usable land either for agriculture or for building lots. Much of what is called conservation today is no more than a useful delaying action. The time may soon come when it will no longer be possible to protest against the despoliation of this or that park, or forest, or river. Hence the conservationist also must face the fact that behind almost every problem of today lies the problem of population. Unless that problem is solved, none of the others can be.

Let us suppose for a moment that those are in the right who say that the context of nature has ceased to be the most desirable context for civilized life, that man can live in a wholly man-made world and that he will in time forget all that he once drew from his contemplation of that world of which he has ceased to be a part. Let us suppose further that his increase in numbers stopped before space itself gave out, and that he has reached what some seem to think of as the ideal state, i.e., living in cities which are almost coextensive with the surface of the earth, nourishing himself on products of laboratories rather than farms, and dealing only with either other men or the machines they have created.

What will he then have become? Will he not have become a creature whose whole being has ceased to resemble Homo sapiens as we in our history have known him? He will have ceased to be consciously a part of that nature from which he sprang. He will no longer have, as he now does, the companionship of other creatures who share with him the mysterious privilege of being alive. The emotions which have inspired a large part of all our literature, music, and art will no longer be meaningful to him. No flower will suggest thoughts too deep for tears. No bird song will remind him of the kind of joy he no longer knows. Will the human race have then become men-like-gods, or only men-like-ants?

To this question the naturalist has his own answer just as he has his own answer to the question why population continues to grow so rapidly in a world already at least beginning to be aware that this growth is a threat. His approach may seem to others somewhat oblique, even distorted by his special interests. But at least his conclusions are the same as those to which many other approaches no less inevitably lead.

The Good Life

André Maurois

Until the outbreak of the First World War in 1914, we might have declared hopefully that mankind seemed to be making some progress toward civilization. The eighteenth century had been called "The Century of Enlightenment" and it is true that science, philosophy, and other matters of the spirit shone brightly during that period. Culture in the eighteenth century, however, was reserved for a numerically small elite; the masses remained steeped in superstition and fanaticism; torture and slavery had not been abolished. The nineteenth century and the beginning of the twentieth saw the rapid advance of democracy and justice. Manners had become gentler. A judicial error could arouse the whole world to indignation, as was seen at the time of the Dreyfus affair. Compassion, patience, and friendship among men seemed to be necessary virtues which the majority tried hard to practice.

In 1961, alas! we must record not further progress but a series of terrible retrogressions. During the course of the Second World War, we witnessed atrocities such as had not been committed for a very long time. Torture had become standard operating procedure rather than an exception lapse. Violence, either by war or mass disorders, had replaced negotiation and compromise. Political partisanship, nationalism, and racial hatred are today resorting to a terrorism that we had a right to believe was obsolete. Our very manners today are more brutal. Our films and our theater reflect a society in which noble sentiment

From *Our Crowded Planet,* edited by Fairfield Osborn (New York, 1962), pp. 215-220. This article was translated by Lawrence G. Blochman.

is rare and a decent way of life held up to ridicule. Our screen is monopolized by murders, holdups, rapes, and orgies. As an indication of how far we have retreated from humanity, compare the motion pictures deriving from the daily life of our times with *La Princesse de Clèves,* a film based on a novel of the seventeenth century.

Why this retrogression? We might have expected the contrary. We might have hoped that democratic education, in giving everyone a chance to share the culture of the past, would have produced a more kindly society. It would seem that the knowledge of the humanities, that heritage common to all men, that vast religious, literary, and artistic patrimony which could unite all peoples in common admiration, should have rendered impossible a certain barbarism which we so aptly called "inhumanity." Actually, the humanities no longer exercise their benevolent influence. For the first time in history men have created the United Nations—and never before in history have men been so disunited. They have drawn up a Declaration of the Rights of Man applicable to all without discrimination because of country, race, or religion—and rarely have the rights of man been so widely disregarded in so many parts of the world. Why?

There are many possible answers to this question because there are underlying reasons of many different kinds which combine to produce the deplorable situation. There is of course the disastrous role played by the two great wars. There are the contributions of false doctrines and the philosophies of despair. And there is another cause, frequently overlooked, which I should like to underline here: *Overpopulation.* You may well ask: What relationship can there be between overpopulation and morals, between overpopulation and the humanities? I shall try to show that the relationship is very close indeed.

First of all, we must understand that the humanities, broadly speaking, are passed on from generation to generation through the media of the family and community. It is not by the philosophers that our children are taught good manners and morals; it is by example. In my biography of Adrienne de La Fayette, I described how girls were brought up in a religious family of the eighteenth century. There was no constraint; the mother appealed to her daughters' reason. But she was an admirable mother and nobody could possible have lived with her without seeking to imitate her. The twelve-year-old of that time wrote letters which in the perfection of style and sentiment could be equaled by few people of our day, even the best and most intelligent.

Now, this familial education, as efficacious as it was solidly based, has disappeared almost entirely today. Our overpopulated cities consist of huge, hastily built barracks in which families that are too big, crowd into apartments that are too small. The children rarely stay home because there is no room for them to play. They go outside to play with the neighbor's children—perhaps to form the nucleus of one of the juvenile gangs which are becoming too nefarious in too many countries today. Education by example is out of style. The bond between generations has grown slack. If I may be permitted to cite a personal example, my own taste for letters was transmitted to me by my highly cultured mother who from my earliest childhood read to me from the great authors, adding her own commentary, so that I should get to know the masterworks of literature at first hand. Very few mothers today would have the chance to exert a similar influence on their offspring.

In the past, familial education was supplemented by community education. The village was a school of neighborliness and friendship. Because a villager knew all his neighbors and also knew that he was destined to spend his life among them, he strove to prove himself a man of good graces and good manners. But in our overpopulated world, the small community is becoming more and more scarce. In African and Asian countries, the law of the village and the law of the tribe have been the sources of all civilization. But since overpopulation and industrialization tend to create urban centers becoming ever more huge, what is to become of the ceremonies and good manners of the past? They are disappearing—or nearly. What has been the basis for our nobler sentiments? Love and affection, which thrive best in the smaller units of society. True, love is not dead, but in our day it assumes its most ephemeral and least lovely aspects. Sexuality will never replace tenderness, nor will it inspire either compassion or tolerance. It is rather akin to brutality.

The salvation of our civilization lies with the humanities as we may learn them from the great books. There is no finer school for fine sentiments than the great novels or the best of the theater. It should be the role of our universities to bring humanity back to its sources. If the universities throughout the world could teach men of all races that in our great poets and philosophers they have friends in common, a strong, invisible web would be woven to join all peoples. But in this domain, too, overpopulation exerts a sorry influence. Competition is bitter in an overpopulated world. Man once produced about as much food as he needed for himself. Today more people are crowded upon a land than the land can feed. Those in excess must learn to earn their living in another manner.

Unfortunately, they stand to earn a more generous living if they study subjects other than the humanities. The modern world, preoccupied especially with machines and armaments, demands engineers, pilots, physicists, chemists, doctors, mechanics. Education is daily becoming more scientific and more technical. If only the student had spare time enough to devote to the humanities! But the techniques of our age have become so complex that to master them a student has no time to spare. The peoples of Asia and Africa are all demanding technicians. The United States, the Soviet Union, the countries of western Europe must turn them out in growing numbers to meet the demand. As a result, the humanities are largely neglected today, and the qualities of style and precise vocabulary are declining in all countries. This creates a dangerous situation. Technicians capable of building powerful instruments of dreadful destruction are no longer able to understand the fine shades of meaning and the delicate distinction between words. They are therefore helpless to defend themselves against the cruel and absolute doctrines which turn their own techniques to the service of fanaticism.

An overpopulated earth will bring forth unintelligent generations because culture demands leisure and silence, which have become lost qualities. Our information media must drop their standards lower and lower in order to reach the level of the new masses, innumerable and unreasoningly exacting. An old Buddhist text predicts: "The time will come when grown men will have the intelligence of a child of ten. These ten-year-old men will be dominated by violent hates, violent malevolence, and a violent desire to kill." We are not far from the day when the murderous mechanism of the nuclear bomb will be placed at the disposition of leaders with infantile minds, leaders hard-pressed by starving multitudes. Should this day come, overpopulation will produce disintegration.

Can this movement be stopped? Can this tide be reversed? Without doubt, it can, but reversal will require great statesmen with the courage to stem the irrational increase of population throughout the world, to re-create within our great, swarming urban centers, the old civilizing cells—the family and the village, fountainheads of all culture, and to restore to favor in all their forms the humanities and the amenities of life. Is all this possible? The history of mankind does record examples of such great reversions. Christianity is one of them.

The wild proliferation of men is a cancer of our planet. A superficial cancer may be cured. But the cure requires treatment—and perseverance.

The Crowded World

Sir Julian Huxley

The Middle Ages were brought to an end by a major revolution in thought and belief, which stimulated the growth of science and the secularization of life at the expense of significance in art and religion, generated the industrial-technological revolution, with its stress on economics and quantitative production at the expense of significance in quality, human values and fulfilment, and culminated in what we are pleased to call the Atomic Age, with two World Wars behind it, the threat of annihilation before it, and an ideological split at its core.

Actually our modern age merits the adjective atomistic rather than atomic. Further, it will soon become very unmodern. For we are on the threshold of major revolution, involving a new pattern of thought and a new approach to human destiny and its practical problems. It will usher in a new phase of human history, which I like to call the Evolutionary Age, because it envisages man as both product and agent of the evolutionary process on this planet.

The new approach is beginning to take effect in two rather distinct fields, of ecology and ideology, and is generating two parallel but linked currents of thought and action, that may be called the Ecological Revolution and the Humanist Revolution.

The population explosion is giving a powerful impetus to both these revolutionary currents. Ecology is the science of relational adjustment—the balanced relations of living organisms with their environment and with each other. It started botanically in a rather modest way as a study of plant communities in different habitats; went on to the

From *Essays of a Humanist*, by Sir Julian Huxley (New York, 1964), pp. 243-250.

fruitful idea of the ecological succession of different plant communities in a given habitat, leading up to an optimum climax community—mixed forest in the humid tropics, rich grassland on the prairies; was extended to take in animal communities, and so to the illuminating concepts of food-chains and adaptive niches; and finally, though rather grudgingly, was still further enlarged to include human as well as biological ecology.

The population explosion has brought us up against a number of tough ecological facts. Man is at last pressing hard on his spatial environment—there is little leeway left for his colonization of new areas of the world's surface. He is pressing hard on his resources, notably non-renewable but also renewable resources. As Professor Harrison Brown has so frighteningly made clear in his book, *The Challenge of Man's Future,* ever-increasing consumption by an ever-increasing number of human beings will lead in a very few generations to the exhaustion of all easily exploitable fossil fuels and high-grade mineral ores, to the taking up of all first-rate agricultural land, and so to the invasion of more and more second-rate marginal land for agriculture. In fact, we are well on our way to ruining our material habitat. But we are beginning to ruin our own spiritual and mental habitat also. Not content with destroying or squandering our resources of material things, we are beginning to destroy the resources of true enjoyment—spiritual, aesthetic, intellectual, emotional. We are spreading great masses of human habitation over the face of the land, neither cities nor suburbs nor towns nor villages, just a vast mass of urban sprawl or subtopia. And to escape from this, people are spilling out farther and farther into the wilder parts and so destroying them. And we are making our cities so big as to be monstrous, so big that they are becoming impossible to live in. Just as there is a maximum possible size for an efficient land animal—you can't have a land animal more than about twice as large as an elephant—so there is a maximum possible efficient size for a city. London, New York, and Tokyo have already got beyond that size.

In spite of all that science and technology can do, world food-production is not keeping up with world population, and the gap between the haves and the have-nots of this world is widening instead of being narrowed.

Meanwhile everywhere, though especially in the so-called Free Enterprise areas of the world, economic practice (and sometimes economic theory) is concerned not primarily with increased production, still less with a truly balanced economy, but with exploitation of resources in the interests of maximized and indiscriminate consumption, even if this involves present waste and future shortage.

Clearly this self-defeating, self-destroying process must be stopped. The population explosion has helped to take our economic blinkers off and has shown us the gross and increasing imbalance between the world's human population and its material resources. Unless we quickly set about achieving some sort of balance between reproduction and production, we shall be dooming our grandchildren and all their descendants, through thousands upon thousands of monotonous generations, to an extremely unpleasant and unsatisfactory existence, overworked and undernourished, overcrowded and unfulfilled.

To stop the process means planned conservation in place of reckless exploitation, regulation and control of human numbers, as well as of industrial and technological enterprise, in place of uninhibited expansion. And this means an ecological approach. Ecology will become the basic science of the new age, with physics and chemistry and technology as its hand-maidens, not its masters. The aim will be to achieve a balanced relation between man and nature, an equilibrium between human needs and world resources.

The Humanist Revolution, on the other hand, is destined to supersede the current pattern of ideas and beliefs about nature (including human nature) and man's place and role in it, with a new vision of reality more in harmony with man's present knowledge and circumstances. This new pattern of ideas can be called humanist, since it is focused on man as a product of natural evolution, not on the vast inanimate cosmos, nor on a God or gods, nor on some unchanging spiritual Absolute. For humanism in this sense, man's duty and destiny is to be the spearhead and creative agent of the overall evolutionary process on this planet.

The explosive growth of scientific and historical knowledge in the past hundred years, especially about biological and human evolution, coupled with the rise of rationalist criticism of established theologies and ancient philosophies, had cleared the ground for this revolution in thought and executed some of the necessary demolition work. But now the population explosion poses the world with the fundamental question of human destiny—*What are people for?* Surely people do not exist just to provide bomb-fodder for an atomic bonfire, or religion-fodder for rival churches, or cannon-fodder for rival nations, or disease-fodder for rival parasites, or labor-fodder for rival economic systems, or ideology-fodder for rival political systems, or even consumer-fodder for profit-making systems. It cannot be their aim just to eat, drink and be merry, and to hell with posterity. Nor merely to prepare for some rather shadowy after-life. It cannot be their destiny to exist in ever larger megalopolitan sprawls, cut off from contact with nature and from the sense of human community and condemned to increasing frustration, noise, mechanical routine, traffic congestion and endless commuting; nor to live out their undernourished lives in some squalid Asian or African village.

When we try to think in more general terms it is clear that the dominant aim of human destiny cannot be anything so banal as just maximum quantity, whether of human beings, machines, works of art, consumer goods, political power, or anything else. Man's dominant aim must be increase in quality—quality of human personality, of achievement, of works of art and craftsmanship, of inner experience, of quality of life and living in general.

"Fulfilment" is probably the embracing word: more fulfilment and less frustration for more human beings. We want more varied and fuller achievement in human societies, as against drabness and shrinkage. We want more variety as against monotony. We want more enjoyment and less suffering. We want more beauty and less ugliness. We want more adventure and disciplined freedom, as against routine and slavishness. We want more knowledge, more interest, more wonder, as against ignorance and apathy.

We want more sense of participation in something enduring and worth while, some embracing project, as against a competitive rat-race, whether with the Russians or our neighbors on the next street. In the most general terms, we want more transcendence of self in the fruitful development of personality: and we want more human dignity not only as against human degradation, but as against more self-imprisonment in the human ego or more escapism. But the inordinate growth of human numbers bars the way to any such desirable revolution, and produces increasing frustration instead of greater fulfilment.

There are many urgent special problems which the population explosion is raising—how to provide the increasing numbers of human beings with their basic quotas of food and shelter, raw materials and energy, health and education, with opportunities for adventure and meditation, for contact with nature and with art, for useful work and fruitful leisure; how to prevent frustration exploding into violence or subsiding into apathy; how to avoid unplanned chaos on the one hand and over-organized authoritarianism on the other.

Behind them all, the long-term general problem remains. Before the human species can settle down to any constructive planning of his future on earth (which, let us remember, is likely to be many times longer than his past, to be reckoned in hundreds of millions of years instead of the hundreds of thousands of his prehistory or the mere millennia of History), it must clear the world's decks for action. If man is not to become the planet's cancer instead of its partner and guide, the threatening plethora of the unborn must be forever banished from the scene.

Above all we need a world population policy—not at some unspecified date in the future, but now. The time has now come to think seriously about population policy. We want every country to have a population policy, just as it has an economic policy or a foreign policy. We want the United Nations to have a population policy. We want all the international agencies of the U.N. to have a population policy.

When I say a population policy, I don't mean that anybody is going to tell every woman how many children she may have, any more than a country which has an economic policy will say how much money an individual businessman is allowed to make and exactly how he should do it. It means that you recognize population as a major problem of national life, that you have a general aim in regard to it, and that you try to devise methods for realizing this aim. And if you have an international population policy, again it doesn't mean dictating to backward countries or anything of that sort; it means not depriving them of the right (which I should assert is a fundamental human right) to scientific information on birth-control, and it means help in regulating and controlling their increase and planning their families.

Its first aim must be to cut down the present excessive rate of increase to manageable proportions: once this is done we can think about planning for an optimum size of world population—which will almost certainly prove to be less than its present total. Meanwhile we, the people of all nations, through the U.N. and its Agencies, through our own national policies and institutions, and through private Foundations, can help those courageous countries which have already launched a population policy of their own, or want to do so, by freely giving advice and assistance and by promoting research on the largest scale.

* * * *

In conclusion I would simply like to go back to where I started and repeat that we must look at the whole question of population increase not merely as an immediate problem to be dealt with *ad hoc*. We must look at it in the light of the new vision of human destiny which human science and learning has revealed to us. We must look at it in the light of the glorious possibilities that are still latent in man, not merely in the light of the obvious fact that the world could be made a little better than it is. We must also look at it in the light of the appalling possibilities for evil and misery that still hang over the future of evolving man.

This vision of the possibilities of wonder and more fruitful fulfilment on the one hand as against frustration and increasing misery and regimentation on the other is the twentieth-century equivalent of the traditional Christian view of salvation as against damnation. I would indeed say that this new point of view that we are reaching, the vision of evolutionary humanism, is essentially a religious one, and that we can and should devote ourselves with truly religious devotion to the cause of ensuring greater fulfilment for the human race in its future destiny. And this involves a furious and concerted attack on the problem of population; for the control of population is, I am quite certain, a prerequisite for any radical improvement in the human lot.

We do indeed need a World Population Policy. We have learnt how to control the forces of outer nature. If we fail to control the forces of our own reproduction, the human race will be sunk in a flood of struggling people, and we, its present representatives, will be conniving at its future disaster.

PART VI

FUTURE PROSPECTS

How Beautiful It Is

It flows out of mystery into mystery: there is no
beginning—
How could there be? And no end—how could there be?
The stars shine in the sky like the spray of a wave
Rushing to meet no shore, and the great music
Blares on forever, but to us very soon
It will be blind. Not we nor our children nor the human
race
Are destined to live forever, the breath will fail,
The eyes will break—perhaps of our own explosive bile
Vented upon each other. Or a stingy peace
Makes parents fools—but far greater witnesses
Will take our places. It is only a little planet
But how beautiful it is.

Robinson Jeffers

From *The Beginning and the End and Other Poems*, by Robinson Jeffers (New York, 1963), p. 29.

The Stationary State

John Stuart Mill

. . . [In] contemplating any progressive movement, not in its nature unlimited, the mind is not satisfied with merely tracing the laws of the movement; it cannot but ask the further question, to what goal? Towards what ultimate point is society tending by its industrial progress? When the progress ceases, in what condition are we to expect that it will leave mankind?

It must always have been seen, more or less distinctly, by political economists, that the increase of wealth is not boundless: that at the end of what they term the progressive state lies the stationary state, that all progress in wealth is but a postponement of this, and that each step in advance is an approach to it. We have now been led to recognize that this ultimate goal is at all times near enough to be fully in view; that we are always on the verge of it, and that if we have not reached it long ago, it is because the goal itself flies before us. The richest and most prosperous countries would very soon attain the stationary state, if no further improvements were made in the productive arts, and if there were a suspension of the overflow of capital from those countries into the uncultivated or ill-cultivated regions of the earth.

This impossibility of ultimately avoiding the stationary state—this irresistible necessity that the stream of human industry should finally spread itself out into an apparently stagnant sea—must have been, to the political economists of the last two generations, an unpleasing and discouraging prospect; for the tone and tendency of their speculations goes completely to identify all that is economically desirable with the

From *Principles of Political Economy*, Vol. II (New York, 1864), pp. 334-340.

progressive state, and with that alone. With [John Ramsay] M'Culloch, for example, prosperity does not mean a large production and a good distribution of wealth, but a rapid increase of it; his test of prosperity is high profits; and as the tendency of that very increase of wealth, which he calls prosperity, is towards low profits, economical progress, according to him, must tend to the extinction of prosperity. Adam Smith always assumes that the condition of the mass of the people, though it may not be positively distressed, must be pinched and stinted in a stationary condition of wealth, and can only be satisfactory in a progressive state. The doctrine that, to however distant a time incessant struggling may put off our doom, the progress of society must "end in shallows and in miseries," far from being, as many people still believe, a wicked invention of Mr. Malthus, was either expressly or tacitly affirmed by his most distinguished predecessors, and can only be successfully combated on his principles. Before attention had been directed to the principle of population as the active force in determining the remuneration of labor, the increase of mankind was virtually treated as a constant quantity: it was, at all events, assumed that in the natural and normal state of human affairs population must constantly increase, from which it followed that a constant increase of the means of support was essential to the physical comfort of the mass of mankind. The publication of Mr. Malthus' Essay is the era from which better views of this subject must be dated; and notwithstanding the acknowledged errors of his first edition, few writers have done more than himself, in the subsequent editions, to promote these juster and more hopeful anticipations.

Even in a progressive state of capital, in old countries, a conscientious or prudential restraint on population is indispensable, to prevent the increase of numbers from outstripping the increase of capital, and the condition of the classes who are at the bottom of society from being deteriorated. Where there is not, in the people, or in some very large proportion of them, a resolute resistance to this deterioration—a determination to preserve an established standard of comfort—the condition of the poorest class sinks, even in a progressive state, to the lowest point which they will consent to endure. The same determination would be equally effectual to keep up their condition in the stationary state, and would be quite as likely to exist. Indeed, even now, the countries in which the greatest prudence is manifested in the regulating of population are often those in which capital increases least rapidly. Where there is an indefinite prospect of employment for increased numbers, there is apt to appear less necessity for prudential restraint. If it were evident that a new hand could not obtain employment but by displacing, or succeeding to, one already employed, the combined influences of prudence and public opinion might in some measure be relied on for restricting the coming generation within the numbers necessary for replacing the present.

I cannot, therefore, regard the stationary state of capital and wealth with the unaffected aversion so generally manifested towards it by political economists of the old school. I am inclined to believe that it would be, on the whole, a very considerable improvement on our present condition. I confess I am not charmed with the ideal of life held out by those who think that the normal state of human beings is that of struggling to get on; that the trampling, crushing, elbowing, and treading on each other's heels, which form the existing type of social life, are the most desirable lot of human kind, or anything but the disagreeable symptoms of one of the phases of industrial progress. The northern and middle states of America are a specimen of this stage of civilization in very favorable circumstances; having, apparently, got rid of all social injustices and inequalities that affect persons of Caucasian race and of the male sex, while the proportion of population to capital and land is such as to ensure abundance to every able-bodied member of the community who does not forfeit it by misconduct. They have the six points of Chartism, and they have no poverty: and all that these advantages seem to have yet done for them (notwithstanding some incipient signs of a better tendency) is that the life of the whole of one sex is devoted to dollar-hunting, and of the other to breeding dollar-hunters. This is not a kind of social perfection which philanthropists to come will feel any very eager desire to assist in realizing. Most fitting, indeed, is it, that while riches are power, and to grow as rich as possible the universal object of ambition, the path to its attainment should be open to all, without favor or partiality. But the best state for human nature is that in which, while no one is poor, no one desires to be richer, nor has any reason to fear being thrust back, by the efforts of others to push themselves forward.

That the energies of mankind should be kept in employment by the struggle for riches, as they were formerly by the struggle of war, until the better minds succeed in educating the others into better things, is undoubtedly more desirable than that they should rust and stagnate. While minds are coarse they require coarse stimuli, and let them have them. In the meantime, those who do not accept the present very early stage of human improvement as its ultimate type, may be excused for being comparatively indifferent to the kind of economical progress which excites the congratulations of ordinary politicians; the mere increase of production and accumulation. For the safety of national independence it is essential that a country should not fall much behind its neighbors in these things. But in themselves they are of little importance, so long as either the increase of population or anything else prevents the mass of the people from reaping any part of the benefit of them. I know not why it should be matter of congratulation that persons who are already richer than any one needs to be, should have doubled their means of consuming things which give little or no pleasure except as representative of wealth; or that numbers of individuals should pass over, every year, from the middle classes into a richer class, or from the class of the occupied rich to that of the unoccupied. It is only in the backward countries of the world that increased production is still an important object: in those most advanced, what is economically needed is a better distribution, of which one indispensable means is a stricter restraint on population. Levelling institutions, either of a just or of an unjust kind, cannot alone accomplish it; they may lower the heights of society, but they cannot, of themselves, permanently raise the depths.

On the other hand, we may suppose this better distribution of property attained, by the joint effect of the prudence and frugality of individuals, and of a system of legislation favoring equality of fortunes, so far as is consistent with the just claim of the individual to the fruits, whether great or small, of his or her own industry. We may suppose, for instance, . . . a limitation of the sum which any one person may acquire by gift or inheritance, to the amount sufficient to constitute a moderate independence. Under this twofold influence, society would exhibit these leading features: a well-paid and affluent body of laborers; no enormous fortunes, except what were earned and accumulated during a single lifetime; but a much larger body of persons than at present, not only exempt from the coarser toils, but with sufficient leisure, both physical and mental, from mechanical details, to cultivate freely the graces of life, and afford examples of them to the classes less favorably circumstanced for their growth. This condition of society, so greatly preferable to the present, is not only perfectly compatible with the stationary state, but, it would seem, more naturally allied with that state than with any other.

There is room in the world, no doubt, and even in old countries, for a great increase of population, supposing the arts of life to go on improving, and capital to increase. But even if innocuous, I confess I see very little reason for desiring it. The density of population necessary to enable mankind to obtain, in the greatest degree, all the advantages both of co-operation and of social intercourse, has, in all the most populous countries, been attained. A population may be too crowded, though all be amply supplied with food and raiment. It is not good for man to be kept perforce at all times in the presence of his species. A world from which solitude is extirpated is a very poor ideal. Solitude, in the sense of being often alone, is essential to any depth of meditation or of character; and solitude in the presence of natural beauty and grandeur is the cradle of thoughts and aspirations which are not only good for the individual, but which society could ill do without. Nor is there much satisfaction in contemplating the world with nothing left to the spontaneous activity of nature; with every rood of land brought into cultivation, which is capable of growing food for human beings; every flowery waste or natural pasture ploughed up, all quadrupeds or birds which are not domesticated for man's use exterminated as his rivals for food, every hedgerow or superfluous tree rooted out, and scarcely a place left where a wild shrub or flower could grow without being eradicated as a weed in the name of improved agriculture. If the earth must lose that great portion of its pleasantness which it owes to things that the unlimited increase of wealth and population would extirpate from it, for the mere purpose of enabling it to support a larger, but not a better or a happier population, I sincerely hope, for the sake of posterity, that they will be content to be stationary, long before necessity compels them to it.

It is scarcely necessary to remark that a stationary condition of capital and population implies no stationary state of human improvement. There would be as much scope as ever for all kinds of mental culture, and moral and social progress; as much room for improving the Art of Living, and much more likelihood of its being improved, when minds ceased to be engrossed by the art of getting on. Even the industrial arts might be as earnestly and as successfully cultivated, with this sole difference, that instead of serving no purpose but the increase of wealth, industrial improvements would produce their legitimate effect, that of abridging labor. Hitherto it is questionable if all the mechanical inventions yet made have lightened the day's toil of any human being. They have enabled a greater population to live the same life of drudgery and imprisonment, and an increased number of manufacturers and others to make fortunes. They have increased the comforts of the middle classes. But they have not yet begun to effect those great changes in human destiny, which it is in their nature and in their futurity to accomplish. Only when, in addition to just institutions, the increase of mankind shall be under the deliberate guidance of judicious foresight, can the conquests made from the powers of nature by the intellect and energy of scientific discoverers, become the common property of the species, and the means of improving and elevating the universal lot.

The Making of an Agitator for Zero

John Fischer

To my astonishment, the political convictions that I had cherished for most of my life have suddenly deserted me. Like my children, these were convictions I loved dearly and had nurtured at considerable expense. When last seen they were—like all of us—somewhat battered by the events of the last decade, but they looked durable enough to last out my time. So I was disconcerted when I found that somehow, during the past winter, they sort of melted away, without my consent and while I was looking somewhere else.

Their place has been usurped by a new set of convictions so radical that they alarm me. If the opposite kind of thing had happened, I would have felt a little melancholy but not surprised, since people traditionally grow more conservative as they get older. But to discover that one has suddenly turned into a militant subversive is downright embarrassing; at times I wonder whether it signals the onset of second childhood.

Except that I seem to be a lot more radical than the children. Those SDS youngsters who go around breaking windows and clubbing policemen now merely depress me with their frivolous irrelevance. So do most other varieties of New Leftists, such as the Women's Liberation movement; if some dire accident should, God forbid, throw one of those ladies into my clutches, she can be sure of instant liberation. I am equally out of tune with those old fogies, the Communists. The differences between capitalism and Communism no longer seem to me worth

From *Harper's*, 240:18-29 (April 1970). Originally published with the title "How I got radicalized: the making of an agitator for Zero."

fighting about, or even arguing, since they are both wrong and beside the point. Or so it seems to me, since the New Vision hit me on my own small road to Damascus.

Let me make it plain that none of this was my doing. I feel as Charles Darwin must have felt during the last leg of his voyage on the *Beagle*. When he embarked he had been a conventional (if slightly lackadaisical) Christian, who took the literal truth of Genesis for granted. He had been raised in that faith, as I was raised a Brass Collar Democrat, and had no thought of forsaking it. Only gradually, while he examined fossil shellfish high in the Andes and measured the growth of coral deposits and the bills of Galapagos finches, did he begin to doubt that the earth and all its inhabitants had been created in six days of October, 4004 B.C., according to the pious calculations of Archbishop James Ussher. By the time he got back to England, he found himself a reluctant evolutionist, soon to be damned as a heretic and underminer of the Established Church. This was not his fault. It was the fault of those damned finches.

Recently I too have been looking at finches, so to speak, although mine are mostly statistical and not nearly as pretty as Darwin's. His gave him a hint about the way the earth's creatures came into being; mine, to my terror, seem to hint at the way they may go out. While I am by no means an uncritical admirer of the human race, I have become rather fond of it, and would hate to see it disappear. Finding ways to save it—if we are not too late already—now strikes me as the political issue which takes precedence over all others.

One of the events which led to my conversion was my unexpected appointment to a committee set up by Governor John Dempsey of Connecticut to work out an environmental policy for our state. Now I had been fretting for quite a while about what is happening to our environment—who hasn't?—but until the work of the committee forced me into systematic study, I had not realized that my political convictions were in danger. Then after looking at certain hairy facts for a few months, I found myself convinced that the Democratic party, and most of our institutions of government, and even the American Way of Life are no damned good. In their present forms, at least, they will have to go. Either that, or everybody goes—and sooner than we think.

To begin with, look at the American Way of Life. Its essence is a belief in growth. Every Chamber of Commerce is bent on making its Podunk grow into the Biggest Little City in the country. Wall Street is dedicated to its search for growth stocks, so that Xerox has become the American ideal—superseding George Washington, who expressed *his* faith in growth by speculating in land. Each year Detroit prays for a bigger car market. Businessmen spend their lives in pursuit of an annual increase in sales, assets, and net profits. All housewives—except for a few slatterns without ambition—yearn for bigger houses, bigger cars, and bigger salary checks. The one national goal that everybody agrees on is an ever-growing Gross National Product. Our modern priesthood—the economists who reassure us that our mystic impulses are moral and holy—recently announced that the GNP would reach a trillion dollars early in this decade. I don't really understand what a trillion is, but when I read the news I rejoiced, along with everybody else. Surely that means that we were in sight of ending poverty, for the first time in human history, so that nobody would ever again need to go hungry or live in a slum.

Now I know better. In these past months I have come to understand that a zooming Gross National Product leads not to salvation, but to suicide. So does a continuing growth in population, highway mileage, kilowatts, plane travel, steel tonnage, or anything else you care to name.

The most important lesson of my life—learned shamefully late—was that nonstop growth just isn't possible, for Americans or anybody else. For we live in what I've learned to recognize as a tight ecological system: a smallish planet with a strictly limited supply of everything, including air, water, and places to dump sewage. There is no conceivable way in which it can be made bigger. If Homo sapiens insists on constant growth, within this system's inelastic walls, something has to pop, or smother. Already the United States is an overpopulated country: not so hopelessly overcrowded as Japan or India, of course, but well beyond the limits which would make a good life attainable for everybody. Stewart Udall, former Secretary of Interior and now a practicing ecologist, has estimated that the optimum population for America would be about 100 million, or half of our present numbers. And unless we do something, drastic and fast, we can expect another 100 million within the next thirty years.

So our prime national goal, I am now convinced, should be to reach Zero Growth Rate as soon as possible. Zero growth in people, in GNP, and in our consumption of everything. That is the only hope of attaining a stable ecology: that is, of halting the deterioration of the environment on which our lives depend.

This of course is a profoundly subversive notion. It runs squarely against the grain of both capitalism and the American dream. It is equally subversive of Communism, since the Communists are just as hooked on the idea of perpetual growth as any American businessman. Indeed, when Khrushchev was top man in the Kremlin, he proclaimed that 1970 would be the year in which the Russians would surpass the

United States in output of goods. They didn't make it: a fact for which their future generations may be grateful, because their environment is just as fragile as ours, and as easily damaged by headlong expansion. If you think the Hudson River and Lake Erie are unique examples of pollution, take a look at the Volga and Lake Baikal.

No political party, here or abroad, has yet even considered adopting Zero Growth Rate as the chief plank in its platform. Neither has any politician dared to speak out loud about what "protection of the environment" really means—although practically all of them seem to have realized, all of a sudden, that it is becoming an issue they can't ignore. So far, most of them have tried to handle it with gingerly platitudes, while keeping their eyes tightly closed to the implications of what they say. In his January State of the Union message, for instance, President Nixon made the customary noises about pollution; but he never even mentioned the population explosion, and he specifically denied that there is any "fundamental contradiction between economic growth and the quality of life." He sounded about as convincing as a doctor telling a cancer patient not to worry about the growth of his tumor.

The Democrats are no better. I have not heard any of them demanding a halt to all immigration, or a steeply progressive income tax on each child beyond two, or an annual bounty to every woman between the ages of fifteen and forty-five who gets through the year without becoming pregnant. Neither Ted Sorensen nor any of the other Kennedy henchmen has yet suggested that a politician with a big family is a spacehog and a hypocrite, unworthy of public trust. No Democrat, to my knowledge, has ever endorsed the views of Dr. René Dubos of Rockefeller University, one of the truly wise men of our time. In an editorial in the November 14, 1969, issue of *Science* he predicted that in order to survive, "mankind will have to develop what might be called a steady state . . . a nearly closed system" in which most materials from tin cans to sewage would be "recycled instead of discarded." His conclusion—that a viable future depends on the creation of "social and economic systems different from the ones in which we live today"—apparently is too radical for any politician I know.

Consequently I feel a little lonesome in my newfound political convictions. The only organization which seems to share them is a tiny one, founded only a few months ago: Zero Population Growth, Inc., with headquarters at 367 State Street, Los Altos, California 94022. Yet I have a hunch that I may not be lonesome for long. Among college students a concern with ecology has become, almost overnight, nearly as popular as sideburns. On many campuses it seems to be succeeding civil rights and Vietnam as The Movement. For example, when the University of Oregon announced last January a new course, "Can Man

Survive?", it drew six thousand students, the biggest class in the university's history. They had to meet in the basketball court because no classroom would hold them.

Who knows? Maybe we agitators for Zero may yet turn out to be the wave of the future.

At the same time I was losing my faith in the virtues of growth, I began to doubt two other articles of the American credo.

One of them is the belief that technology can fix anything. Like most of us, I had always taken it for granted that any problem could be solved if we just applied enough science, money, and good old American know-how. Is the world's population outrunning its food supply? Well, then, let's put the laboratories to work inventing high-yield strains of rice and wheat, better fertilizers, ways to harvest seaweed, hydroponic methods for growing food without soil. If the air is becoming unbreathable, surely the technologists can find ways to clean it up. If our transportation system is a national disgrace, all we have to do is call in the miracle men who built a shuttle service to the moon; certainly they should be able to figure out some way to get a train from New York to New Haven on time.

I was in East Haddam, Connecticut, looking at an atomic power plant, when I began to suspect that technology might not be the answer after all. While I can't go along with the young Luddites who have decided that science is evil and that all inventions since the wheel ought to be destroyed, I am persuaded that technology is a servant of only limited usefulness, and highly unreliable. When it does solve a problem, it often creates two new ones—and their side effects are usually hard to foresee.

One of the things that brought me to East Haddam was curiosity about the automobile. Since the gasoline engine is the main polluter of the air, maybe it should be replaced with some kind of electric motor? That of course would require an immense increase in our production of electric power, in order to recharge ten million batteries every night. Where would it come from? Virtually all waterpower sites already are in use. More coal- and oil-fired power stations don't sound like a good idea, since they too pour smoke into the atmosphere—and coal mining already has ruined countless streams and hundreds of thousands of acres of irreplaceable land. Atomic power, then?

At first glance, the East Haddam plant, which is fairly typical of the new technology, looked encouraging. It is not as painful an eyesore as coal-burning stations, and not a wisp of smoke was in sight. When I began to ask questions, however, the company's public-relations man admitted that there are a few little problems. For one thing, the plant's innards are cooled with water pumped out of the Connecticut River. When it flows back in, this water raises the river's temperature by about twenty degrees, for a considerable distance. Apparently this has not yet done any serious damage to the shad, the only fish kept under careful surveillance; but its effect on other fish and algae, fish eggs, micro-organisms, and the general ecology of the river is substantial though still unmeasured.

It would be possible, though expensive, for the company to build cooling towers, where the water would trickle over a series of baffles before returning to the river. In the process it would lose its heat to the atmosphere. But this, in turn, threatens climatic changes, such as banks of artificial fog rolling eastward over Long Island Sound, and serious wastage of water through evaporation from a river system where water already is in precarious supply. Moreover, neither this process nor any other now known would eliminate the slight, but not negligible, radiation which every atomic plant throws off, nor the remote but still omni-present chance of a nuclear accident which could take thousands of lives. The building of an additional twenty plants along the banks of the Connecticut—which some estimates call for, in order to meet future demand for electricity—would be a clear invitation to an ecological disaster.

In the end I began to suspect that there is no harmless way to meet the demands for power of a rising population, with rising living standards—much less for a new herd of millions of electric cars. Every additional kilowatt levies some tax upon the environment, in one form or another. The Fourth Law of Thermodynamics seems to be: "There is no free lunch."

Every time you look at one of the marvels of modern technology, you find a by-product—unintended, unpredictable, and often lethal. Since World War II American agriculture has performed miracles in increasing production. One result was that we were able for years to send a shipload of free wheat every day to India, saving millions from starvation. The by-products were: (1) a steady rise in India's population; (2) the poisoning of our streams and lakes with insecticides and chemical fertilizers; (3) the forced migration of some ten million people from the countryside to city slums, as agriculture became so efficient it no longer needed their labor.

Again, the jet plane is an unquestionable convenience, capable of whisking a New Yorker, say, to either the French Riviera or Southern California in a tenth of the time he could travel by ship or car, and at lower cost. But when he reaches his destination, the passenger finds the beaches coated with oil (intended to fuel planes, if it hadn't spilled) and the air thick with smog (thanks in good part to the jets, each of which spews out as much hydrocarbon as ten thousand automobiles.

Moreover, technology works best on things nobody really needs, such as collecting moon rocks or building supersonic transport planes. Whenever we try to apply it to something serious, it usually falls on its face.

An obvious case in point is the railroads. We already have the technology to build fast, comfortable passenger trains. Such trains are, in fact, already in operation in Japan, Italy, and a few other countries. Experimental samples—the Metroliners and Turbotrains—also are running with spectacular success between Washington and Boston. If we had enough of them to handle commuter and middle-distance traffic throughout the country, we could stop building the highways and airports which disfigure our countryside, reduce the number of automobiles contaminating the air, and solve many problems of urban congestion. But so far we have not been able to apply the relatively simple technology needed to accomplish these aims, because some tough political decisions have to be made before we can unleash the scientists and engineers. We would have to divert to the railroads many of the billions in subsidy which we now lavish on highways and air routes. We would have to get rid of our present railway management—in general, the most incompetent in American industry—and retire the doddering old codgers of the Railway Brotherhoods who make such a mess out of running our trains. This might mean public ownership of a good many rail lines. It certainly would mean all-out war with the unions, the auto and aviation industries, and the highway lobby. It would mean ruthless application of the No Growth principle to roads, cars, and planes, while we make sensible use instead of something we already have: some 20,000 miles of railways.

All this requires political action, of the most radical kind. Until our Great Slob Society is willing to take it, technology is helpless.

My final apostasy from the American Creed was loss of faith in private property. I am now persuaded that there no longer is such a thing as truly private property, at least in land. That was a luxury we could afford only when the continent was sparsely settled. Today the use a man makes of his land cannot be left to his private decision alone, since eventually it is bound to affect everybody else. This conclusion I reached in anguish, since I own a tiny patch of land and value its privacy above anything money can buy.

What radicalized me on this score was the Department of Agriculture and Dr. Ian McHarg. From those dull volumes of statistics which the Department publishes from time to time, I discovered that usable land is fast becoming a scarce resource—and that we are wasting it with an almost criminal lack of foresight. Every year, more than a million acres of farm and forest land is being eaten up by highways, airports, reservoirs, and real-estate developments. The best, too, in most cases, since the rich, flat bottom lands are the most tempting to developers.

Since America is, for the moment, producing a surplus of many crops, this destruction of farmland has not yet caused much public alarm. But some day, not too far off, the rising curve of population and the falling curve of food-growing land inevitably are going to intersect. That is the day when we may begin to understand what hunger means.

Long before that, however, we may be gasping for breath. For green plants are our only source of oxygen. They also are the great purifiers of the atmosphere, since in the process of photosynthesis they absorb carbon dioxide—an assignment which gets harder every day, as our chimneys and exhaust pipes spew out ever-bigger tonnage of carbon gases. This is a function not only of trees and grass, but also of the tiny microorganisms in the sea. Indeed, its phytoplankton produces some 70 percent of all the oxygen on which life depends. These are delicate little creatures, easily killed by the sewage, chemicals, and oil wastes which already are contaminating every ocean in the world. Nobody knows when the scale will tip: when there are no longer enough green growing things to preserve the finely balanced mixture of gases in the atmosphere, by absorbing carbon dioxide and generating oxygen. All we know is that man is pressing down hard on the lethal end of the scale.

The Survivable Society, if we are able to construct it, will no longer permit a farmer to convert his meadow into a parking lot any time he likes. He will have to understand that his quick profit may, quite literally, take the bread out of his grandchildren's mouths, and the oxygen from their lungs. For the same reasons, housing developments will not be located where they suit the whim of a real-estate speculator or even the convenience of the residents. They will have to go on those few carefully chosen sites where they will do the least damage to the landscape, and to the life-giving greenery which it supports.

This is one of the lessons taught by Ian McHarg in his extraordinary book, *Design With Nature,* recently published by Natural History Press. Alas, its price, $19.95, will keep it from reaching the people who need it most. It ought to be excerpted into a pocket-size volume—entitled, perhaps, "The Thoughts of McHarg"—and distributed free in every school and supermarket.

The current excitement about the environment will not come to much, I am afraid, unless it radicalizes millions of Americans. The conservative ideas put forth by President Nixon—spending a few billion for sewage-treatment plants and abatement of air pollution—will not even begin to create the Survivable Society. That can be brought about only by radical political action—radical enough to change the whole structure of government, the economy, and our national goals.

How the Survivable State will work is something I cannot guess; its design is a job for the coming generation of political scientists. The radical vision can, however, give us a glimpse of what it might look like. It will measure every new law, every dollar of investment by a cardinal yardstick: Will this help us accomplish a zero rate of growth and a stabilized environment? It will be skeptical of technology, including those inventions which purport to help clean up our earthly mess. Accordingly it will have an Anti-Patent Office, which will forbid the use of any technological discovery until the Office figures out fairly precisely what its side effects might be. (If they can't be foreseen, then the invention goes into deep freeze.) The use of land, water, and air will not be left to private decision, since their preservation will be recognized as a public trust. The landlord whose incinerator smokes will be pilloried; the tanker skipper who flushes his oil tanks at sea will be hanged at the nearest yardarm for the capital crime of oxygen destruction. On the other hand, the gardener will stand at the top of the social hierarchy, and the citizen who razes a supermarket and plants its acreage in trees will be proclaimed a Hero of the Republic. I won't live to see the day, of course; but I hope somebody will.

The Challenge to Renewal

Lewis Mumford

The age that we live in threatens worldwide catastrophe; but it likewise holds forth unexpected hope and unexampled promise. Ours is no time for faint-hearted men. No matter how rugged the obstacles that confront us, we must push on, like Bunyan's Pilgrim, not heeding the Worldly Wisemen who are torpid to the danger and fearful of the promise. If we do not sink into the Slough of Despond, we may yet find our way to the Delectable Mountains and to that fair land where the sun shines night and day. The shadows that now fall across our path measure the height we have still to climb.

Perhaps never before have the peoples of the world been so close to losing the very core of their humanity; for of what use are cosmic energies, if they are handled by disoriented and demoralized men? But the very threat of general disintegration has also increased the possibility for a rapid and radical improvement in the condition of man. The most generous dreams of the past have now become immediate practical necessities: a worldwide co-operation of peoples, a more just distribution of all the goods of life; the use of knowledge and energy for the service of life, and the use of life itself for the extension of the human spirit to provinces where human values and purposes could not heretofore penetrate. If we awaken in time to overcome the automatisms and irrational compulsions that are now pushing the nations toward destruction, we shall create a universal community. Even if we awaken only belatedly, the fresh insight and the new philosophy that

From *The Conduct of Life*, by Lewis Mumford (New York, 1951), pp. 3-4, 268-273.

might have saved us in the first place will be needed to carry us through the dark days ahead.

The renewal of life is the burden and challenge of our time: its urgency lightens its risks and its difficulties. For the first time in history, the tribes and nations have the means of entering into an active partnership, as wide and unrestricted as the planet itself. Universal fellowship, which the higher religions conceived for many millenniums as mankind's destiny, now has become technically feasible as well as ideally conceivable: to seal that promise with acts of political and economic co-operation on a worldwide scale has likewise become a practical imperative.

Nothing short of such a transformation will keep the human race from sliding back still further into barbarism: a barbarism whose powers of destruction have been multiplied by the very scientific knowledge that most modern thinkers, up to our own age, believed a sure guarantee of the continued advance of civilization. The rational conduct of life, plainly, demands something far different from the automatic extension of science and invention.

* * * *

The Swiss historian Jacob Burckhardt predicted that the corruptions and weaknesses already observable in Western civilization by the middle of the nineteenth century would result in the coming of the Terrible Simplifiers: people who, with ruthless decision and unstinted force, would overthrow even the good institutions that were, in fact, stifling the growth of the human spirit. "People," he wrote, "may not yet like to imagine a world whose rulers completely ignore law, prosperity, profitable labor, and industry, credit, etc.," a world governed by military corporations and single parties: but such a world becomes possible when the majority no longer, through orderly means, exercises the initiative in continuously re-forming and re-directing institutions to serve human purposes. What the virtuous will not do in a reasonable constructive way, the criminal and barbarian take upon themselves to do, negatively and irrationally, for the sheer pleasure of destruction. When individuals shun responsibility as persons, their place is taken politically by a tyrant, who recovers freedom of initiative through crime.

Even before Burckhardt, Dostoyevsky had predicted, with remarkable prescience, what would occur. In that enigmatic narrative, Letters from the Underworld, in which Dostoyevsky put so many challenging truths into the mouth of his sniveling, repulsive chief character, the veritable prototype of Hitler, he described the utilitarian heaven of the nineteenth century: the heaven, still, of popular current science, in which all the questions that had heretofore troubled men would be precisely answered and all human acts would be mathematically computed according to nature's laws, so that the world will cease to know any wrongdoing. Then he observes: "I should not be surprised if, amid all this order and regularity of the future, there should suddenly arise, from some quarter or another, some gentleman of lowborn—or rather, of retrograde and cynical—demeanor, who, setting his arms akimbo, should say to you all: 'How now, gentlemen? Would it not be a good thing if, with one accord, we were to kick all this solemn wisdom to the winds and send these logarithms to the devil, and to begin to live our lives again according to our own stupid whimsy?'"

This is the nihilistic answer to the serious condition that every civilization at length finds itself in: the result of over-organization, the multiplication of superfluous wants, an excess of regularity and routine in the conduct of daily life, a fossilization of even happy rituals: all resulting in a failure of human initiative and a dull submission to what seems an overbearing impersonal determinism. In such an existence people eat for the sake of supporting meat packers' organizations and dairymen's associations, they guard their health carefully for the sake of creating dividends for their life insurance corporations, they earn their daily living for the sake of paying dues, taxes, mortgages, installments on their car or their television sets, or fulfilling their quota in a Five-Year Plan: in short, they satisfy the essential needs of life in general, the preparatory acts deplete the appropriate consummations.

Such a society as ours eventually ties itself up into knots by its inability to put first things first. When a community reaches a point where no one can make a decision of the simplest sort without bringing into play an elaborate technique of research or accountancy, without enlisting the aid of innumerable specialists who take responsibility for only their minute fragments of the process, all the normal acts of living must be slowed down to such an extent that the economies originally achieved by division of labor and large-scale organization are nullified. Thus the technique of diagnosis becomes as burdensome to the patient as his disease: indeed it becomes an auxiliary disease. At that point, the life of a community will be stalled and frustrated: it will not be capable of anticipating or circumventing the simplest crisis.

But no community can permit itself to be stalled for long; for if we do not find a benign method of simplification, then the Terrible Simplifiers will come on the scene, recapturing freedom through savagery and charlatanism, if not through the polite forms sanctioned by an over-developed civilization. When our apparatus of fact-finding and truth-proving becomes too complicated, the Terrible Simplifiers will resort to brazen lies and childish superstitions. If our factual historians ostracize the Burckhardts and the Henry Adamses for daring to look into the future on the basis of their knowledge and wisdom, people who seek guidance will take to astrological horoscopes as a substitute.

To escape the Terrible Simplifiers one must recognize the actual danger of the condition through which they obtain their ascendance over the frustrated majority: for the condition these charlatans profess to correct is in fact a serious one. Instead of closing our eyes to its existence, we must use art and reason to effect a benign simplification, which will give back authority to the human person. Life belongs to the free-living and mobile creatures, not to the encrusted ones; and to restore the initiative to life and participate in its renewal, we must counterbalance every fresh complexity, every mechanical refinement, every increase in quantitative goods or quantitative knowledge, every advance in manipulative technique, every threat of superabundance or surfeit, with stricter habits of evaluation, rejection, choice. To achieve that capacity we must consciously resist every kind of automatism: buy nothing merely because it is advertised, use no invention merely because it has been put on the market, follow no practice merely because it is fashionable. We must approach every part of our lives with the spirit in which Thoreau undertook his housekeeping at Walden Pond: be ready, like him, even to throw out a simple stone, if it proves too much trouble to dust. Otherwise, the sheer quantitative increase in the data of scientific knowledge will produce ignorance: and the constant increase in goods will produce a poverty of life.

There is no domain today where methods of simplification must not be introduced. Because of the uninhibited production of books and scholarly reviews, there is, for example, hardly a single province of thought where the human mind can make an adequate survey of the literature on any subject, except of the minutest province, come to intelligent conclusions, or move confidently from reflection to practice. Our ingenious mechanical methods of solving this problem, like the invention of the microfilm, increase the size of the total burden: the only true salvation, in this and every other sphere, is *voluntarily to restrict production at its source and to increase our selectivity:* both true simplifications, though only the enlightened and the courageous can apply them. This holds for the whole routine of life: never to use mechanical power when human muscles can conveniently do the work, never to use a motor car where one might easily walk, never to acquire information or knowledge except for the satisfaction of some immediate or prospective want—such modes of simplification, though individually insignificant, add up to a considerable degree of emancipation. A popular mentor, himself no enemy of the profit motive, once suggested that one should never waste time opening second class mail; and if that advice were generally taken, at least in America, a vast amount of time and energy would be saved: indeed whole forests would be preserved. Many other institutions will, in time, follow the example of a progressive school in New York: a school that once gave all its students intelligence tests and heaped up a vast mass of unused and unnecessary data. Now it has destroyed these files and it gives special tests only to those who gravely need such additional checks.

In Western countries one of the prime marks of an organic change in our culture—the hallmark of a new brotherhood and sisterhood—would probably be the drastic reduction of the now compulsive habits of smoking and drinking: along with this would go a return, on the part of women, to a mode of wearing their hair which would forego the elaborate mechanical or chemical procedure for producing fashionable uniformity of curl after original Hollywood models. Hundreds of thousands of acres of land would be freed for food-growing by curtailment of tobacco alone, along with some slight direct improvement of health, and a release, if the movement were spontaneous, from neurotic obsessions. The fact that even in a time of worldwide starvation, after the Second World War, no one dared to suggest even a partial conversion of tobacco land to food-growing, shows how rigid, rigid almost to the point of *rigor mortis*, our civilization has become: with no sufficient power of adaptation to reality. Nor is this demonstration lessened by the fact that there is record of starving men asking for tobacco ahead of food: that merely shows the depth of our perversion of life-needs.

Many effective kinds of simplification will perhaps be resisted at first on the ground that this means a "lowering of standards." But this overlooks the fact that many of our standards are themselves extraneous and purposeless. What is lowered from the standpoint of mechanical complexity or social prestige may be raised from the standpoint of the vital function served, as when the offices of friendship themselves replace, as Emerson advocated in his essay on household economy, elaborate preparations of food and service, of napery and silver.

Consider the kind of frugal peasant living that Rousseau first advocated, when he chose to live in a simple cottage, instead of in the mansion of his patron, surrounded by "comforts": all this wipes away time-consuming rituals and costly temptations to indigestion. Or consider the gain in physical freedom modern woman made, when the corset and petticoats, the breast-deformers, pelvis-constrictors, backbone-curvers of the Victorian period gave way to the garb of the early 1920's, without girdle, brassière, or even stocking supporters: a high point of freedom in clothes from which women sheepishly recoiled under the deft browbeating of manufacturers with something to sell.

Naturally the sort of simplification needed must itself conform to life-standards. Thoreau's over-simplification of his diet, for instance, probably undermined his constitution and gave encouragement to the tuberculosis from which he finally died. By now we know that a diet consisting of a single kind of food is not part of nature's economy: the amino acids appear to nourish the body only when various ones are present in different kinds of food: so that the lesson of life is not to confuse simplicity with monotony. So, too, a tap of running water, fed by gravity from a distant spring, is in the long run a far more simple device, judged by the total man-hours used in production and service, than the daily fetching of water in a bucket: as the bucket, in turn, is more simple than making even more frequent journeys to the spring to slake one's thirst directly. Simplicity does not avoid mechanical aids: it seeks only not to be victimized by them. That image should save us from the imbecile simplifiers, who reckon simplicity, not in terms of its total result on living, but in terms of immediate first costs or in a pious lack of visible apparatus.

Sporadically, during the last three centuries, many benign simplifications have in fact come to pass throughout Western civilization; though, as in the case of women's dress just noted, they have sometimes been followed by reactions that have left us as badly off as ever. Rousseau, coming after the Quakers, carried their simplification of manners through to diet, to child nurture, and to education; while Hahnemann began a similar change in medicine, a change followed through by Dr. William Osler, under whom hundreds of spurious drugs and complicated prescriptions were discarded, in favor of the Hippocratic attention to diet and rest and natural restoratives. In handicraft and art and architecture the same general change was effected, first by William Morris, in his rule: "Possess nothing that you do not know to be useful or believe to be beautiful." Modern architecture, though it has often been distracted and perverted by technical over-elaboration, can justify its essential innovations as an attempt to simplify the background of living, so that the poorest member of our society will have as orderly and harmonious an environment as the richest: it has discarded complicated forms as a badge of class and conspicuous waste. Wherever the machine is intelligently adapted to human needs, it has the effect of simplifying the routine of life and releasing the human agent from slavish mechanical tasks. It is only where the person abdicates that mechanization presents a threat.

But in order to recover initiative for the person, we must go over our whole routine of life, as with a surgeon's knife, to eliminate every element of purposeless materialism, to cut every binding of too-neat red tape, to remove the fatty tissue that imposes extra burdens on our organs and slows down all our vital processes. Simplicity itself is not the aim of this effort: no, the purpose is to use simplicity to promote spontaneity and freedom, so that we may do justice to life's new occasions and singular moments. For what Ruskin said of the difference between a great painter, like Tintoretto, and a low painter, like Teniers, holds for every manifestation of life: the inferior painter, not recognizing the difference between high and low, between what is intensely moving and what is emotionally inert, gives every part of his painting the same refinement of finish, the same care of detail. The great painter, on the other hand, knows that life is too short to treat every part of it with equal care: so he concentrates on the passages of maximum significance and treats hastily, even contemptuously, the minor passages: his short-cuts and simplifications are an effort to give a better account of what matters. This reduction to essentials is the main art of life.

Once Upon a Time —A Fable of Student Power

Neil Postman

Once upon a time in the City of New York civilized life very nearly came to an end. The streets were covered with dirt, and there was no one to tidy them. The air and rivers were polluted, and no one could cleanse them. The schools were rundown, and no one believed in them. Each day brought a new strike, and each strike brought new hardships. Crime and strife and disorder and rudeness were to be found everywhere. The young fought the old. The workers fought the students. The whites fought the blacks. The city was bankrupt.

When things came to their most desperate moment, the City Fathers met to consider the problem. But they could suggest no cures, for their morale was very low and their imagination dulled by hatred and confusion. There was nothing for the Mayor to do but to declare a state of emergency. He had done this before during snowstorms and power failures, but now he felt even more justified. "Our city," he said, "is under siege, like the ancient cities of Jericho and Troy. But *our* enemies are sloth and poverty and indifference and hatred." As you can see, he was a very wise Mayor, but not so wise as to say exactly how these enemies could be dispersed. Thus, though a state of emergency officially existed, neither the Mayor nor anyone else could think of anything to do that would make their situation better rather than worse. And then an extraordinary thing happened.

One of the Mayor's aides, knowing full well what the future held for the city, had decided to flee with his family to the country. In order

From the *New York Times Magazine*, IV:10 (June 14, 1970).

to prepare himself for his exodus to a strange environment, he began to read Henry David Thoreau's "Walden," which he had been told was a useful handbook on how to survive in the country. While reading the book, he came upon the following passage:

"Students should not play life, or study it merely, while the community supports them at this expensive game, but earnestly live it from the beginning to end. How could youths better learn to live than by at once trying the experiment of living?"

The aide sensed immediately that he was in the presence of an exceedingly good idea. And he sought an audience with the Mayor. He showed the passage to the Mayor, who was extremely depressed and in no mood to read from books, since he had already scoured books of lore and wisdom in search of help but had found nothing. "What does it mean?" said the Mayor angrily. The aide replied: "Nothing less than a way to our salvation."

He then explained to the Mayor that the students in the public schools had heretofore been part of the general problem whereas, with some slight imagination and a change of perspective, they might easily become part of the general solution. He pointed out that from junior high school on up to senior high school, there were approximately 400,000 able-bodied, energetic young men and women who could be used as a resource to make the city livable again. "But how can we use them?" asked the Mayor. "And what would happen to their education if we did?"

To this the aide replied, "They will find their education in the process of saving their city. And as for their lessons in school, we have ample evidence that the young do not exactly appreciate them and are even now turning against their teachers and their schools." The aide, who had come armed with statistics (as aides are wont to do), pointed out that the city was spending $5-million a year merely replacing broken school windows and that almost one-third of all the students enrolled in the schools did not even show up on any given day. "Yes, I know," said the Mayor sadly. "Woe unto us." Wrong," said the aide brashly. "The boredom and destructiveness and pent-up energy that are now an affliction to us can be turned to our advantage."

The Mayor was not quite convinced, but having no better idea of his own he appointed his aide Chairman of the Emergency Education Committee, and the aide at once made plans to remove almost 400,000 students from their dreary classrooms and their even drearier lessons, so that their energy and talents might be used to repair the desecrated environment.

When these plans became known, there was a great hue and cry against them, for people in distress will sometimes prefer a problem that is familiar to a solution that is not. For instance, the teachers complained that their contract contained no provision for such unusual procedures. To this the aide replied that the *spirit* of their contract compelled them to help educate our youth, and that education can take many forms and be conducted in many places. "It is not written in any holy book," he observed, "that an education must occur in a small room with chairs in it."

Some parents complained that the plan was un-American and that its compulsory nature was hateful to them. To this the aide replied that the plan was based on the practices of earlier Americans who required their young to assist in controlling the environment in order to insure the survival of the group. "Our schools," he added, "have never hesitated to compel. The question is not, nor has it ever been, to compel or not to compel, but rather, which things ought to be compelled."

And even some children complained, although not many. They said that their God-given right to spend 12 years of their lives, at public expense, sitting in a classroom was being trampled. To this complaint the aide replied that they were confusing a luxury with a right, and that, in any case, the community could no longer afford either. "Besides," he added, "of all the God-given rights man has identified, none takes precedence over his right to survive."

And so, the curriculum of the public schools of New York City became known as Operation Survival, and all the children from 7th grade through 12th grade became part of it. Here are some of the things that they were obliged to do:

On Monday morning of every week, 400,000 children had to help clean up their own neighborhoods. They swept the streets, canned the garbage, removed the litter from empty lots, and hosed the dust and graffiti from the pavements and walls. Wednesday mornings were reserved for beautifying the city. Students planted trees and flowers, tended the grass and shrubs, painted subway stations and other eyesores, and even repaired broken-down public buildings, starting with their own schools.

Each day, 5,000 students (mostly juniors and seniors in high school) were given responsibility to direct traffic on the city streets, so that all the policemen who previously had done this were freed to keep a sharp eye out for criminals. Each day, 5,000 students were asked to help deliver the mail, so that it soon became possible to have mail delivered twice a day—as it had been done in days of yore.

Several thousand students were also used to establish and maintain day-care centers, so that young mothers, many on welfare, were free to find gainful employment. Each student was also assigned to meet with two elementary-school students on Tuesday and Thursday afternoons to teach them to read, to write, and to do arithmetic. Twenty thousand students were asked to substitute, on one afternoon a week, for certain adults whose jobs the students could perform without injury or loss of efficiency. These adults were then free to attend school or, if they preferred, to assist the students in their efforts to save their city.

The students were also assigned to publish a newspaper in every neighborhood of the city, in which they were able to include much information that good citizens need to have. Students organized science fairs, block parties and rock festivals, and they formed, in every neighborhood, both an orchestra and a theater company. Some students assisted in hospitals, helped to register voters, and produced radio and television programs which were aired on city stations. There was still time to hold a year-round City Olympics in which every child competed in some sport or other.

It came to pass, as you might expect, that the college students in the city yearned to participate in the general plan, and thus another 100,000 young people became available to serve the community. The college students ran a "jitney" service from the residential boroughs to Manhattan and back. Using their own cars and partly subsidized by the city, the students quickly established a kind of auxiliary, semi-public transportation system, which reduced the number of cars coming into Manhattan, took some of the load off the subways, and diminished air pollution—in one stroke.

College students were empowered to give parking and litter tickets, thus freeing policemen more than ever for real detective work. They were permitted to organize seminars, film festivals, and arrange lectures for junior and senior high school students; and on a U.H.F. television channel, set aside for the purpose, they gave advanced courses in a variety of subjects every day from 3 P.M. to 10 P.M. They also helped to organize and run drug-addiction rehabilitation centers, and they launched campaigns to inform people of their legal rights, their nutritional needs, and of available medical facilities.

Because this is a fable and not a fairy tale, it cannot be said that all the problems of the city were solved. But several extraordinary things did happen. The city began to come alive and its citizens found new reason to hope that they could save themselves. Young people who had been alienated from their environment assumed a proprietary interest in it. Older people who had regarded the young as unruly and parasitic came to respect them. There followed from this a revival of courtesy and a diminution of crime, for there was less reason than before to be angry at one's neighbors and to wish to assault them.

Amazingly, most of the students found that while they did not "receive" an education, they were able to create a quite adequate one. They lived, each day, their social studies and geography and communication and biology and many other things that decent and proper people know about, including the belief that everyone must share equally in creating a livable city, no matter what he or she becomes later on. It even came to pass that the older people, being guided by the example of the young, took a renewed interest in restoring their environment and, at the very least, refused to participate in its destruction.

Now, it would be foolish to deny that there were not certain problems attending this whole adventure. For instance, there were thousands of children who would otherwise have known the principal rivers of Uruguay who had to live out their lives in ignorance of these facts. There were hundred of teachers who felt their training had been wasted because they could not educate children unless it were done in a classroom. As you can imagine, it was also exceedingly difficult to grade students on their activities, and after a while, almost all tests ceased. This made many people unhappy, for many reasons, but most of all because no one could tell the dumb children from the smart children anymore.

But the Mayor, who was, after all, a very shrewd politician, promised that as soon as the emergency was over everything would be restored to normal. Meanwhile, everybody lived happily ever after—in a state of emergency, but quite able to cope with it.

Pollution and the Public

Frank M. Potter, Jr.

Pollution is limited by neither internal nor external political boundaries. Dirty air and water pass easily from country to country, and people downwind and downstream can only suffer, possibly comforted to know that their hands are no cleaner than those of their neighbors. We are challenged to develop new ways of attacking pollution. In so doing, we must take account of the deficiencies built into the system and, wherever possible, should adapt corrective techniques to the situation as we find it, not as we would have it.

The first and basic need is for a more authoritative information-gathering network, and methods of getting that information at minimum cost to those who need it. This need affects all institutions at all levels. It is being met only superficially at the present time. How it should be carried out and who should do it are important questions not yet resolved: bureaucracy makes strong arguments for keeping the apparatus out of governmental hands, and the profit motive provides strong reasons for keeping it out of the marketplace.

In developing any such information network, emphasis must be placed upon the excellence of the service—differences of opinion are no vice when responsible and adequately documented, and unanimity of opinion ought to be cause for concern.

Unfortunately, the time scale within which we must respond to environmental challenges is so compressed that whatever information

From *The Center Magazine*, III:18-23 (May 1970).

and control systems we can develop may still be unable to operate effectively. The rate of technological change will probably remain rapid, although, as suggested by John Platt in his article "What We Must Do" (*Science*, November 28, 1969), a leveling off is likely in some areas. The objective thus becomes to develop sufficiently responsive systems to permit society to react to new crises before they have acquired unstoppable momentum.

These difficulties are compounded by our inaccurate and inadequate trouble-sensing procedures. We do not seem able to react even when problems are foreseen; we respond only when they have become massive and less easily managed. Inaction in turn requires far greater corrective force than would have been necessary had we reacted sooner and more adequately.

This also points up the failings of the more or less simplistic solutions that we tend to adopt as a means of correcting environmental problems, which are rarely if ever simple in origin, and are not usually curable by the simple solutions presented to and accepted by those who make the crucial decisions.

Finally, we have never seriously set out to define what we mean when we talk about an "optimum" or "livable" environment. True, we all tend to make these judgments on a subjective, non-analytical basis, and we focus on issues with which we may personally and emotionally be involved. The tennis-shoed little old lady may grieve for the Redwoods or a threatened brook without realizing that bigger and more serious problems may threaten much more basic values—perhaps life itself.

Subjective judgments on these questions are unavoidable, and may not be undesirable. But at the same time it would seem important to devote a portion of our energies to an informed effort to *define* the public interest and clarify some of the conflicts that are inevitably involved. If, for example, we continue to favor the internal combustion engine as an integral element of our transportation system, what will this mean in terms of projected levels of air pollution, climate, and human health? Should we not, in other words, develop a base line, from which we may then judge the consequences and costs of proposed new courses of action?

The traditional approach to the development of social control systems has involved the creation of regulatory agencies acting as expert arbiters to protect the public interest. This approach has been spectacularly unsuccessful: the regulators have inevitably become captives of the very industries they were established to regulate. Consciously or not, the regulators have adopted roles as promoters and protectors of the theoretically regulated, and leave little hope that improved environmental protection would result from the establishment of a new super-regulatory agency.

A newer method suggested for controlling rampant environmental degradation involves establishing technical and technological monitoring systems; that is, putting scientists in the position of active maintenance, control, and dissemination of environmental information and protective measures. But this effort is hardly more likely to succeed, since it requires a degree of political sensitivity and aggressiveness foreign and perhaps even antithetical to the scientific method, and certainly inconsistent with history and current practice.

The most adequate solution appears to lie in putting necessary information into the hands of the concerned public, which has the most direct interest, and by giving it better tools and ways of calling environmental miscreants to account. Of course we cannot prevent the bureaucrat or the entrepreneur from making decisions which have short-term advantages for him but long-term advantages for us, but we *can* require him to make his decisions and reasons public, and to provide a forum to review those decisions with broad social interests in mind.

In effect, this would involve building into the decision-making structure of government the ability and desire to consider long-term and ecological consequences of activities, a process that might be accomplished through a number of specific steps:

•Long-term effects of programs and policies must be examined and detailed as a matter of public record.

•Procedures must be established to follow up government programs and projects, determining whether the environmental effects were those anticipated, and if not, why not. (Here again the public should be given easy access to the full record, and procedures should be established permitting citizens to put the appropriate agencies on the spot.)

•Executive agencies should be required as a matter of procedure to obtain the views of other interested federal, state, or local groups, public and private, on questions related to their programs. Responsible criticisms should be answered on the record, and if no answer is forthcoming, or if the answer is unsatisfactory, procedures should be established to permit judicial review.

•Public agencies, in adopting specific programs, should also be required to show how these programs are best adapted to the total needs of the situation. Where reasonable alternatives exist, these should be described, and an explanation given as to why they were not adopted.

•Each agency taking action should be legally required to justify why any action at all was desirable. This is not so simpleminded as it sounds: the Corps of Engineers is hard put to defend itself when asked to develop cost/benefit calculations for *not* building a dam. Assembling a group of technologists and/or engineers presupposes great pressure to do *something*, and the option of not going forward at all is often obscured or ignored.

This last requirement suggests itself for nongovernmental areas of endeavor as well. Highway builders, land developers, and others have a far easier job in making their cases than do their opponents. A heavy burden of proof is placed upon the people who presume to speak for the public interest. To get into court they must show that active harm will result, not balanced by the putative good provided by the proposed activity. The burden is misplaced—those who wish to use environmental assets should be required to show that the balance *favors* their proposals.

•We also need mechanisms for more rapid, extensive, and convenient public review of major public and private agency decisions. Perhaps a Public Defender for the Environment, with authority to review general governmental policies and pass upon specific problems considered to have significant environmental consequences might be a solution. In extraordinary cases, this Defender could be given the authority to issue temporary cease-and-desist orders to prevent the otherwise inevitable destruction of important resources. Control procedures must be set up to prevent such a Defender from acting irresponsibly, or to force him to act in proper cases.

•We must encourage the public to participate more effectively in the making of decisions with environmental implications, on which it has no presently measurable impact. This means citizen action programs, keyed to the issues of the day. Call them lobbies, pressure groups, or anything else; color them important. Their actions should be coordinated to have a meaningful impact upon the legislative bodies whose decisions affect us all.

•We also need to develop new ways of funding citizen organizations with environmental objectives. Where they act to protect common assets, they should be supported by the public treasury or by the organizations whose actions created the problem.

Greater citizen participation might be accomplished by the enactment of a federal statute to the effect that any person or group winning or perhaps even instituting a court case based upon the violation of a federal pollution law should be entitled, in the discretion of the court, to recover reasonable fees and costs. It would be necessary to spell out in detail the type of cases in which such relief would be appropriate, but the basic idea merits discussion.

•In many ways it would appear more desirable to force the would-be polluter himself to underwrite the costs of protecting the resources that he has threatened. This could be done by requiring a public bond to be filed by agencies which propose to take actions with potentially undesirable environmental consequences. That bond would be subject to forfeiture if an anti-pollution law were violated or if unforeseen environmental consequences should occur, and the funds might be applied to legal fees or to cleaning up the resultant damages.

•We should also step up our efforts to find more adequate technological solutions to the problems which technology has created. The most effective and least harmful method developed to clean up the Santa Barbara oil spill was the massive use of straw, men, and hand rakes—hardly a creative response. Transferring oil from Alaska's North Slope to world markets may create serious environmental threats: the use of gigantic ice-breaking tankers endangers the Arctic Ocean, and the use of overland pipelines threatens a tundra that has remained substantially unchanged for many, many years. Both techniques menace a fragile ecology that might take centuries to recover if something unforeseen should happen.

It is almost inconceivable that more effective and less expensive techniques could not be found to meet these and other environmental hazards of the time. The civilization that put men on the moon ought to be able to do better.

In 1968, several congressmen formed an unofficial Ad-Hoc Committee on the Environment as a channel for communication on environmental issues between the Congress and interested scientists and informed citizens. The committee now numbers a hundred and twenty, and is in regular contact with a hundred and thirty-two expert advisers. Membership on the committee is open to any interested legislator, senator or representative, Republican or Democrat. This step does not entirely satisfy the need for better information, but it seems to be a long step in the right direction. The information network available to members of that committee may soon be expanded to meet state and local demands for better environmental information, and ought also to be useful to other groups with similar concerns.

Legislation has been considered in the Congress which could go far toward arming citizens' organizations with better information on what federal agencies are doing and why they are doing it. The National Environmental Policy Act of 1969, sponsored by Senator Henry Jackson and Representative John Dingell, contains language to this effect, as does the airport construction bill recently passed by the House. It remains to be seen, of course, to what extent the executive agencies will be successful in their inevitable efforts to weaken the impact of these measures, though their jobs will be made more difficult by the certain knowledge that interested legislators will be watching.

These steps and the ones that remain to be taken are hopeful signs in an area in which hope is uncommon. If anything, these efforts should be accelerated; we may not be able to afford more delay, and we should begin to exercise what talents we have for imaginative and bold departures from the patterns of behavior no longer adequate to our needs.

The international community is also rapidly becoming aware of the dangers of environmental degradation. Sweden has taken an important step by proposing a worldwide Conference on the Environment for 1972, under the auspices of the United Nations. The hazard is one which many nations recognize, but this recognition must be tempered by the realization that agreement is easy in principle but not in fact. Everyone is against pollution, but the ranks of enthusiasts thin quickly as specific problems arise and specific remedies are proposed.

We have been less than successful in dealing with environmental problems on the local and national level. Internationally, our record is even worse. The history of the international fishing and whaling commissions does not encourage a sanguine view of the future. The United Nations, in turn, has neither the constituency nor the commitment to resolve foreseeable international environmental conflicts. It was not created for this purpose, and would require extensive internal change if it were to take them up seriously.

The need for better information channels is as great internationally as it is on a smaller scale. If anything, political solutions are more easily blocked than at state and national levels, and no one has yet devised a workable system of sanctions to minimize those problems, which all concede do exist.

If it is true that the interests of small groups are often at odds with those of the large societies in which they exist, how much more true is this of nations, whose antagonisms are more easily created and sustained, and whose common concerns may be deliberately obscured? Downwind and downstream nations from those applying persistent pesticides may see their own problems clearly, but their apprehensions are likely to be viewed as quite unimportant by the nation creating the problem. That nation may well consider its first interest to be protection of the health and food supply of its own citizens, and look upon undesirable side effects as someone else's problem. Unfortunately, they may be everyone else's problem.

The strongest peaceful sanction we have available to influence international decisions appears to be public opinion. More attention might profitably be devoted to the use of public disclosure as a stimulus for more adequate decisions about the international environmental issues. A weak reed it may be, but it must serve until we can find a stronger substitute.

One simple illustration of how such pressures might work can be seen in proposed treaties for the use of the seas. This crucial area has been perceived clearly by national interests as a vast potential source of food and mineral resources, and consequently as critical to their survival.

We must pass over without further analysis the critical issue of sanctions as beyond the scope of this article and as beyond the ability of the concerned parties to resolve at this time. We shall also assume, for the purpose of argument, that it will eventually become possible to develop working treaty relationships with the affected nations and that such a treaty will provide an operating structure as well as a policy-making body.

What suggestions may be made to provide some assurance that the vast assets of the ocean will be used for the common good, and not misused on behalf of narrow segments of humanity? Proposals have been made to provide a focus for scientific impact at the policy-making level; these disciplines will of necessity be represented at the operating levels as well. The proposals do not appear to be entirely adequate to current needs—they will be as inadequate to solve international issues as they are to solve problems on the national scale.

We need an Ombudsman for the Seas.

The functions of such an Ombudsman would be simple: to review and to comment upon proposed actions by the operating arm of the treaty organization and others, to consult with the policy-making arm on matters which are or which ought to be under consideration, and to make recommendations to these and to all nations on ways to use, without misusing, the oceans.

This latter point is particularly important, since the seas could be affected by the activities of nations which may not be treaty signatories—even by nations which are entirely landlocked. Inland rivers and estuaries play an important role in the life cycles of fish and other species important to man, and are in turn highly vulnerable to actions affecting airsheds or watersheds with oceanic outlets. Few nations in the world remain entirely oblivious to the opinions of others, and the ability of the Ombudsman to focus worldwide attention upon previously ignored problems could develop into a highly valuable tool.

As sanctions are developed for the international treaty, consideration should also be given to making sanction available to the Ombudsman, under adequate control procedures. The Ombudsman should not be a policeman: there will be enough problems without adding new ones. However, there should be a close working relationship with whatever organization handles the operations of the treaty organization.

The Ombudsman should have direct access to current oceanographic and ecological information about the seas. Again it would be desirable to keep informational and experimental activities separated from their primary functions; it would also be important to keep them separated from the conventional channels of authority within the operating arm of the treaty organization.

History indicates that, in the seas as elsewhere, strong pressures will be brought to bear by those seeking to exploit these resources. It will be critically important to build into the treaty organization some form of countervailing pressures to ensure that the long-term productivity of the oceans is not endangered by man's effort to turn these assets to limited advantage. If we have learned nothing else from the ecologists, we know now that we exist within a closed system and that we must develop processes and procedures that will permit us to recycle those resources that we must use. To this end, the Ombudsman can serve us well.

For a number of reasons it would seem desirable to create a three- or five-member organization of Ombudsmen with staggered, rotating memberships, and a semi-permanent professional staff. Continuity is important, but a constant access to fresh blood provides a responsiveness to challenge that will be invaluable.

A highly structured decision-making apparatus within the organization itself may not be desirable. No member should be given a power of veto; indeed, if any member sees a particular problem as important, and his colleagues do not share his views, he should still be given latitude to study the problem and to report on it to the appropriate bodies, supporting his report with whatever evidence is available.

The Ombudsmen should be required to submit an annual report on their operations to the treaty organization, and copies of this report should be given wide distribution to member nations as well as to the United Nations. Dissenting views should be made available in the same form. The incentive to review specific problems might come from within the organization itself, or it might come from any member nations. If review is declined, the reasons for disapproval ought to be spelled out in detail.

Funding is critical. As one of the important functions of the treaty, the Ombudsmen should be assured of a regular budget, subject to no diminution for political reasons. Unless the organization can be free of budgetary apprehensions, its work must inevitably suffer.

Clearly the problems of protection of the global environment are not confined to its oceans. Treaties for the oceans are only a beginning —but there is no good reason why these treaties should not be viewed as the first real steps toward more comprehensive and adequate environmental protection. Men require a world that men can live in.

The oceans are important for a number of reasons. It has been shown that they are not as productive as they were once thought to be, in terms of long-term food sources for humanity. We cannot carelessly develop the oceans as a habitat or dumping grounds, but must concern ourselves with protecting this vital element of Spaceship Earth. At the same time, we may perhaps take a halting step toward developing techniques that may prove effective in other areas as well.

The environmental outlook is not encouraging. We are coming to recognize that, however distant the prospect, we have degraded the environment in which we live and on which we depend, and that the quality of our lives—very likely our existence itself—is in danger.

Paradoxically, there is really no villain at whom one can point the finger of blame—unless we are all villains. However, a strong case can be made that the real problem is that our social institutions have proved inadequate to carry the burdens suddenly thrust upon them. Our ability to manipulate the physical world has far outstripped the social institutions and protective devices that might otherwise have shielded us.

Therefore, our new struggle must be to achieve more direct and intimate participation in the decision-making process by citizens and broadly based interest groups. This, coupled with fuller disclosure of the process itself, promises significant benefits.

Representatives of the bureaucratic/industrial complex may say that such changes in the making of decisions will slow those decisions down and will make the process itself more cumbersome. They would be quite correct: new projects and proposals will be hampered and new enterprise will be slowed down.

There may be extraordinary occasions in which such delay cannot be tolerated, but I believe that more careful and balanced consideration of the consequences of future enterprise will benefit society. Once upon a time, social policy was designed to encourage new forms of commercial, industrial, and governmental activity: the elaborate fiction of the corporation was devised to permit men to act collectively without risk to their personal fortunes. America encouraged the growth of the maritime industry, then the railroads, and we are still encouraging an aircraft industry which scarcely needs encouragement and which may be serving interests opposed to those of society (who really needs the SST, anyway?).

Perhaps it is time to cut back this encouragement—to build up the other side of the case: the protection and nurture of the citizen—the human being. New procedures to protect the earth's endangered life-support system may indeed produce a less vigorous exploitation of her resources—but consider the alternatives.

Criteria for an Optimum Human Environment

Hugh H. Iltis, Orie L. Loucks, and Peter Andrews

Almost every current issue of the major science journals contains evidence of an overwhelming interest in one urgent question: Shall a single species of animal, man, be permitted to dominate the earth so that life, as we know it, is threatened? The uniformity of the theme is significant but if there is consensus, it is only as to the need for concern. Each discipline looks differently at the problem of what to do about man's imminent potential to modify the earth through environmental control. Proposals to study ways of directing present trends in population, space and resource relationships toward an "optimum" for man are so diverse as to bewilder both scientists and the national granting agencies.

Arrogance toward Nature

It is no thirst for argument that compels us to add a further view. Rather it is the sad recognition of major deficiencies in policies guiding support of research on the restoration of the quality of our environment. Many of us find the present situation so desperate that even short-term treatments of the symptoms look attractive. We rapidly lose sight of man's recent origins, probably on the high African plains and the natural environment that shaped him. Part of the scientific community also accepts what Lynn White has called our Judeo-Christian arrogance toward nature, and is gambling that our superior technology will deliver the necessary food, clean water and fresh air. But are these the only necessities? Few research proposals effectively ask whether man has other

From the *Bulletin of the Atomic Scientists*, 26:2-6 (January 1970).

than these basic needs, or whether there is a limit to the artificiality of the environment that he can tolerate.

In addition, we wish to examine which disciplines have the responsibility to initiate and carry out the research needed to reveal the limits of man's tolerance to environmental modification and control. We are especially concerned that there is, on the one hand, an unfortunate conviction that social criteria for environmental quality can have no innate biological basis—that they are only conventions. Yet, on the other hand, there is increasing evidence suggesting that mental health and the emotional stability of populations may be profoundly influenced by frustrating aspects of an urban, biologically artificial environment.

There have been numerous proposals for large-scale inter-disciplinary studies of our environment and of the future of man, but such studies must have sufficient breadth to treat conflicting views and to seek to reconcile them. We know of no proposal that would combine the research capabilities of a group studying environmental design with those of a group examining the psychological and mental health responses of man to natural landscapes. The annual mass migration of city man into natural landscapes which provide diversity is a matter of concern to the social scientist, whose research will only be fully satisfactory when joined with studies that quantify the landscape quality, the psychology of individual human response, and the evolutionary basis of man's possible genetic adaptations to nature. The following summary of recent work may provide a basis for scientists in all areas to seek and support even greater breadth in our studies of present and future environments for man.

"Web of Life"

Two major theses are sufficiently well established to provide the positive foundation of our argument. First, we believe the inter-dependency of organisms, popularly known as the "web of life," is essential to maintaining life and a natural environment as we know it. The suffocation of aquatic life in water systems, and the spread of pollutants in the air and on the land, make it clear that the "web of life" for many major ecosystems is seriously threatened. The abrupt extinction of otherwise incidental organisms, or their depletion to the point of no return, threatens permanently to impair our fresh water systems and coastlines, as well as the vegetation of urban regions.

Second, man's recent evolution is now well enough understood for it to play a major part in elucidating the total relation of man to his natural environment. The major selection stresses operating on man's physical evolution have also had some meaning for the development of social structures. These must be considered together with the immense potential of learned adaptations over the entire geologic period of this physical evolution. Unfortunately, scientists, like most of us moderns, are city dwellers dependent on social conventions, and so have become progressively more and more isolated from the landscape where man developed, and where the benchmarks pointing to man's survival may now be found. They, of all men, must recognize that drastic environmental manipulations by modern man must be examined as part of a continuing evolutionary sequence.

The immediacy of problems relating to environmental control is so startling that the threat of a frightening and unwanted future is another point of departure for our views. At the present rate of advance in technology and agriculture, with an unabated expansion of population, it will be only a few years until all of life, even in the atmosphere and the oceans, will be under the conscious dictates of man. While this general result must be accepted by all of us as inevitable, the methods leading to its control offer some flexibility. It is among these that we must weigh and reweigh the cost-benefit ratios, not only for the next 25 or 50 years, but for the next 25,000 years or more. The increasing scope of the threat to man's existence within this controlled environment demands radically new criteria for judging "benefits to man" and "optimum environments."

It would be perverse not to acknowledge the immense debt of modern man to technological development. In mastering his environment, man has been permitted a cultural explosion and attendant intricate civilization made possible by the very inventiveness of modern agriculture, an inventiveness which must not falter if the world is to feed even its present population. Agricultural technology of the nineteenth and twentieth centuries, from Liebig and the gasoline engine to hybrid corn, weed killers and pesticides, has broken an exploitative barrier leading to greatly increased production and prosperity in favored regions of the world. But this very success has imposed upon man an even greater responsibility for managing all of his physical and biotic environment to his best and sustained advantage.

The view also has been expressed recently that the "balance of nature," upset by massive use of non-disintegrating detergents and pesticides, will be restored by "new engineering." Such a view is necessarily based on the assumption that it is only an engineering problem to provide "an environment [for man] relatively free from unwanted man-produced stress." But when the engineering is successful, the very success dissipates our abilities to see the human being as part of a complex biological balance. The more successful technology and agriculture become, the more difficult it is to ask pertinent questions and to expect sensible answers on the long-range stability of the system we build.

The Right Questions?

Inspired by recent success, some chemical and agricultural authorities still hold firmly that we can feed the world by using suitable means to increase productivity, and there is a conviction that we can and must bend all of nature to our human will. But if open space were known to be as important to man as is food, would we not find ways to assure both? Who among us has such confidence in modern science and technology that he is satisfied we know enough, or that we are even asking the right questions, to ensure our survival beyond the current technological assault upon our environment. The optimism of post-World War II days that man can solve his problems—the faith in science that we of Western culture learn almost as infants—appears more and more unfounded.

To answer "what does man now need?" we must ask "where has he come from?" and "what evidence is there of continuing genetic ties to surroundings similar to those of his past?"

Theodosius Dobzhansky and others have stressed that man is indeed unique, but we cannot overlook the fact that the uniqueness does not separate him from animals. Man is the product of over a hundred million years of evolution among mammals, over 45 million years among primates, and over 15 million years among apes. While this morphology has been essentially human for about two million years, the most refined neurological and physical attributes are perhaps but a few hundred thousand years old.

Selection and Adaptation

G. G. Simpson notes that those among our primate ancestors with faulty senses, who misjudged distances when jumping for a tree branch or who didn't hear the approach of predators, died. Only those with the agility and alertness that permitted survival in ruthless nature lived to contribute to our present-day gene pool. Such selection pressure continued with little modification until the rise of effective medical treatment and social reforms during the last five generations. In the modern artificial environment it is easy to forget the implications of selection and adaptation. George Schaller points out in "The Year of the Gorilla" that the gorilla behaves in the zoo as a dangerous and erratic brute. But in his natural environment in the tropical forests of Africa, he is shy, mild, alert and well-coordinated. Neither gorilla nor man can be fully investigated without considering the environments to which he is adapted.

Unique as we may think we are, it seems likely that we are genetically programmed to a natural habitat of clean air and a varied green landscape, like any other mammal. To be relaxed and feel healthy usually means simply allowing our bodies to react as evolution has equipped them to do for 100 million years. Physically and genetically we appear best adapted to a tropical savanna, but as a civilized animal we adapt culturally to cities and towns. For scores of centuries in the temperate zones we have tried to imitate in our houses not only the climate but the setting of our evolutionary past: warm humid air, green plants, and even animal companions. Today those of us who can afford it may even build a greenhouse or swimming pool next to our living room, buy a place in the country, or at least take our children vacationing at the seashore. The specific physiological reactions to natural beauty and diversity, to the shapes and color of nature, especially to green, to the motions and sounds of other animals, we do not comprehend and are reluctant to include in studies of environmental quality. Yet it is evident that nature in our daily lives must be thought of, not as a luxury to be made available if possible, but as part of our inherent indispensable biological need. It must be included in studies of resource policies for man.

Dependence on Nature

Studies in anthropology, psychology, ethology and environmental design have obvious implications for our attempts to structure a biologically sound human environment. Unfortunately, these results frequently are masked by the specifics of the studies themselves. Except for some pioneer work by Konrad Lorenz followed up at several symposia in Europe, nothing has been done to systematize these studies or extend their implications to modern social and economic planning. For example, Robert Ardrey's popular work, "The Territorial Imperative," explores territoriality as a basic animal attribute, and tries to extend it to man. But his evidence is somewhat limited, and we have no clear conception of what the thwarting of this instinct does to decrease human happiness. The more extensive studies on the nature of aggression explore the genetic roots of animal conflicts, roots that were slowly developed by natural selection over millions of generations. These studies suggest that the sources of drive, achievement, and even of conflict within the family and war among men are likely to be related to primitive animal responses as well as to culture.

Evidence exists that man is genetically adapted to a nomadic hunting life, living in small family groups and having only rare contact with larger groups. As such he led a precarious day-to-day existence, with strong selective removal due to competition with other animals, including other groups of humans. Such was the population structure to which man was ecologically restricted and adapted until as recently as 500 generations ago. Unless there has since been a shift in the major causes of human mortality before the breeding age (and except for resistance to specific diseases there is no such evidence), this period is far too short for any significant changes to have occurred in man's genetic makeup.

Studies of neuro-physiological responses to many characteristics of the environment are also an essential part of investigating genetic dependence on natural as opposed to artificial environment. The rapidly expanding work on electroencephalography in relation to stimuli is providing evidence of a need for frequent change in the environment for at least short periods, or, more specifically, for qualities of diversity in it. There is reason to believe that the electrical rhythms in the brain are highly responsive to changes in surroundings when these take the full attention of the subject. The rise of mechanisms for maintaining constant attention to the surroundings can be seen clearly as a product of long-term selection pressures in a "hunter and hunted" environment. Conversely, a monotonous environment produces wave patterns contributing to fatigue. One wonders what the stimuli of brick and asphalt jungles, or the monotony of corn fields, do to the nervous system. Biotic as well as cultural diversity from the neurological point of view, may well be fundamental to the general health that figures prominently in the discussions of environmental quality.

Results with Patients

The interesting results of Maxwell Weismann in taking chronically hospitalized mental patients camping are also worth noting. Hiking through the woods was the most cherished activity. Some 35 of the 90 patients were returned to their communities within three months after the two-week camping experience. Other studies have shown similar results. Many considerations are involved, but it seems possible that in a person whose cultural load has twisted normal functioning into bizarre reactions, his innate genetic drives still continue to function. Responses attuned to natural adaptations would require no conscious effort. An equally plausible interpretation of Weismann's results is that the direct stimuli of the out-of-doors, of nature alone, produces a response toward the more normal. A definitive investigation of the bases for these responses is needed as guidance to urban planners and public health specialists.

These examples are concerned with the negative effects which many see as resulting from the unnatural qualities of man's present, mostly urban, environment. Aldous Huxley ventures a further opinion as he considers the abnormal adaptation of those hopeless victims of mental illness who appear most normal: "These millions of abnormally normal people, living without fuss in a society to which, if they were fully human beings, they ought not to be adjusted, still cherish 'the illusion of individuality,' but in fact they have been to a great extent de-individualized. Their conformity is developing into something like uniformity. But uniformity and freedom are incompatible. Uniformity and mental health are incompatible as well. . . . Man is not made to be an automaton, and if he becomes one, the basis for mental health is lost."

Clearly, a program of research could tell us more about man's subtle genetic dependence on the environment of his evolution. But of one thing we can be sure: only from study of human behavior in its evolutionary context can we investigate the influence of the environment on the life and fate of modern man. Even now we can see the bases by which to judge quality in our environment, if we are to maintain some semblance of one which is biologically optimum for humans.

We do not plead for a return to nature, but for re-examination of how to use science and technology to create environments for human living. While sociological betterment of the environment can do much to relieve poverty and misery, the argument that an expanding economy and increased material wealth alone would produce a Utopia is now substantially discounted. Instead, a natural concern for the quality of life in our affluent society is evident. But few economists or scientists have tried to identify the major elements of the quality we seek, and no one at all has attempted to use evolutionary principles in the search for quality. Solutions to the problems raised by attempts to evaluate quality will not be found before there is tentative agreement on the bases for judging an optimum human environment. A large body of evidence from studies in evolution, medicine, psychology, sociology, and anthropology suggests clearly that such an environment will be a compromise between one in which humans have maximum contact with the properties of the environment to which they are innately adapted, and a more urban environment in which learned adaptations and social conventions are relied upon to overcome primitive needs.

Our option to choose a balance between these two extremes runs out very soon. Awareness of the urgency to do something is national, and initial responses may be noted in several well-established but relatively narrow scientific disciplines. There has been the recent revival of eugenics. A balanced view has been proposed by Leonard Ornstein (*Bulletin*, June 1967), who agrees with others that positive improvements in man's genetic make-up must wait until we are vastly more knowledgeable. He recommends control of degenerating effects from uncontrolled mutation (in the absence of high selection) until more positive measures can be taken.

An "Impossible" Challenge

More extreme views have been expressed that man could be changed genetically to fit any future, but the means to do this and the moral justification of the aims sought are still far from being resolved. Many support the so-called evolutionary and technological optimists who, unlike their forefathers of little more than a generation ago, believe man can be changed radically when the time comes. They show a faith that science has proved its ability to draw on an expanding technology to do the impossible. The technologically impossible seems to have been accomplished time and time again during the past two or three generations, and may happen again. But some important scientific objectives have not been achieved, and we are likely to become more aware of the failures of science, of the truly impossible, as the irreversible disruptions of highly complex biological systems become more evident.

We suggest that the alternative to genetic modification of man is to select a course where the objectives only verge on the impossible. Let us regard the study and documentation of criteria for an environmental optimum as the "impossible" challenge for science and technology in the next two decades! Although considerable research in biology, sociology, and environmental design is already directed to this objective, there are several other types of study required that we outline briefly, simply to indicate the scope of the challenge.

First, a thorough examination must be undertaken of the extent to which man's evolutionary heritage dominates his activity both as an individual and in groups. The survival advantage of certain group activities has clearly figured in his evolutionary success and adaptive culture. Although cultural adaptation now dominates the biological in the evolution of man, his basic animal nature has not changed. Research leading to adequate understanding of the need to meet innate genetic demands lies in the field of biology, and more specifically in a combination of genetics, physical anthropology and ethology.

Second, we need to understand more of how cultural adaptations and social conventions of man permit him to succeed in an artificial environment. Cultural adaptation is the basis of his success as a gregarious social animal, and it will continue to be the basis by which he modifies evolutionarily imposed adaptations. Medical studies suggest there may be a limit to the magnitude of cultural adaptations, and that for some people this is nearly reached. Studies in sociology, cultural anthropology and psychology are all necessary to such research, in combination with environmental design and quantitative analysis of diversity in the native landscape.

Third, relationships between the health of individuals, both mental and physical, and the properties of the environment in which they live should be a fundamental area of research. It is easy to forget that we should expect as much genetic variability in the capacity of individuals to adjust to artificial environments as we find in the physical characteristics of man. Some portions of the population should be expected to have a greater inherent commitment to the natural environment, and will react strongly if deprived of it. Others may be much more neutral. Studies of the population as a whole must take into account the variability in reaction, and must therefore consider population genetics as well as psychiatry and environmental design.

Fourth, environmental qualities should be programed so as to optimize for the maximal expression of evolutionary (i.e. human) capabilities at the weakest link in the ontogenic development of human needs. While there are many critical periods during our life, we believe the ties to natural environments to be most vital during youth. We have abundant evidence on our campuses and in our cities that the dislodgement of youth presents one—if not the most—serious obstacle to successful adoption of more complex social structures. The dislodgement of man in an artificial environment will vary throughout his ontogeny. Even the small child or infant cannot be expected to be indifferent to changes in the gross characteristics of his community, as he cannot within his own family.

Young men and women accept many of the modern social conventions, but retain the highly questioning mind that once led to new and better ways to hunt and forage. By early middle age, man's physical and mental agility has changed and he becomes a stronger adherent to the social conventions that make his own society possible. During the rise of modern man on the high African plains, and continuing into modern primitive societies, each community was very much dependent on its young men. They contributed to hunting and community protection through their strength and agility, commodities for which there is declining demand in modern society. Survival in the primitive groups was to some degree dependent on the willingness of youth to innovate and take risks, and this has become a fixed adaptation, requiring outlets of expression.

Over 30 years ago, sociologist W. F. Ogburn suggested that society in the future would require "prolonging infancy to, say, thirty or forty years or even longer." Is not our 20-year educational sequence a poorly-veiled attempt to do just that? From an evolutionary point of view will not this dislodgement of youth present the most serious obstacle to successful adoption of more complex social structures? We are compelled to acknowledge that our over-all technological environment for youth has not compensated for the loss of the challenges of the hunt and the freedom of the Veldt. The disruptions on our campuses and in the cities indicates the need to plan environmental optima for this weakest link in the human need for expression of evolutionary capabilities.

Finally, systems ecology is developing the capacity for considering all of the relationships and their interactions simultaneously. The notion of fully describing the optimum for any organism may seem presumptuous. It requires measurement of every type of response, particularly behavioral responses, and their statement as a series of component equations. Synthesis in the form of a complex model permits mathematical examination of an optimum for the system as a whole. Until recently it seemed more reasonable to study such optimization for important resources such as fisheries, but the capability is available and relevant to the study of the environmental optimum of man, and its application must now be pursued vigorously.

These five approaches to the study of human environment provide an objective base for investigating the environmental optimum for man. We cannot close this discussion, however, without pointing out that the final decision, both as to the choice of the optimum and its implementation, is an ethical one. There is an optimum for the sick, and another for the well; there is an optimum for the maladjusted, and another for the well-adjusted. But in treating the problems of the poor and minority groups, in our preoccupation with their immediate relief, we may continue to overlook the ways in which cultural demands of the modern, sub-optimum environment go far beyond the capacity of learned adaptations.

A Compromise?

Considering our scientific effort to learn the functions and structure of the human body, and of the physical environment around us, the limited knowledge of man's relationships to his environment is appalling. Because of the very success of our scientific establishment we are faced with population densities and environmental contaminants that have left us no alternative but to undertake control of the environment itself. In this undertaking let us understand the need to choose a humane compromise—a balance between the evolutionary demands we cannot deny except with great emotional and physical misery, and the fruits of an unbelievably varied civilization we are loath to give up.

Yet are we even considering such a compromise? With rare exceptions are we not continuing to destroy much that remains of man's natural environment with little thought for the profit of the remote future? In the conflict between preservationists and industrialists (or agriculturalists) the latter have had it their way, standing as they do for "progress" and "modern living." While the balance between these conflicts is slowly changing, preservationists continue to be regarded as sentimentalists rather than realists.

Theodosius Dobzhansky says that "the preponderance of cultural over biological evolution will continue to increase in the foreseeable future." We could not wish this to be otherwise; adaptation to the environment by culture is more rapid and efficient than biological adaptation. But social structures cannot continue indefinitely to become more complex and further removed from evolutionary forces. At some stage a compromise must be reached with man's innate evolutionary adaptability.

Need for Continuing Study

We believe that the evidence of man's need for nature, particularly its diversity, is sufficient to justify a determined effort by the scientific community to obtain definitive answers to the questions we have posed. The techniques for studying the problems are to be found in separate disciplines, and there is a sufficient measure of willingness among scientists to undertake the new approaches. But the first steps will be faltering and financial support will be slow in coming.

Now that buttercups are rare, at least symbolically, and springs often silent, why study them? Have there not already been several generations for whom the fields and woods are nearly a closed book? We could encourage the book to close forever, and we might succeed, but in so doing we might fail disastrously. The desire to see and smell and know has not yet been suppressed and enthusiasm for natural history continues to bring vitality to millions. Let us recognize that we are a product of evolution, without apology for the close affinities with our primate forebears. We need only prepare consciously to make a compromise between our cultural and our genetic heritage by striking a balance of social structures with maintenance of natural environments. Most important, we must discover the mechanisms of environmental influence on man. There is no other satisfactory approach to an optimum environment.

The Unbelievable Future

René Dubos

Rebels in Search of a Cause

This book should have been written in anger. I should be expressing in the strongest possible terms my anguish at seeing so many human and natural values spoiled or destroyed in affluent societies, as well as my indignation at the failure of the scientific community to organize a systematic effort against the desecration of life and nature. Environmental ugliness and the rape of nature can be forgiven when they result from poverty, but not when they occur in the midst of plenty and indeed are produced by wealth. The neglect of human problems by the scientific establishment might be justified if it were due to lack of resources or of methods of approach, but cannot be forgiven in a society which can always find enough money to deal with the issues that concern selfish interests.

Unfortunately, writing in anger requires talents I do not possess. This is my excuse for presenting instead a mild discussion of our collective guilt.

We claim that human relationships and communion with nature are the ultimate sources of happiness and beauty. Yet we do not hesitate to spoil our surroundings and human associations for the sake of efficiency in acquiring power and wealth. Our collective sense of guilt comes from a general awareness that our praise of human and natural values is hypocrisy as long as we practice social indifference and convert our land into a gigantic dump.

From *So Human an Animal*, by René Dubos (New York, 1969), pp. 3-30.

Phrases like "one world" and the "brotherhood of man" occur endlessly in conversations and official discourses at the very time that political wars and race riots are raging all over the world. Politicians and real-estate operators advocate programs for the beautification of cities and highways, while allowing the exciting grandeur of the American wilderness to degenerate into an immense ugliness. Brush is overgrowing mountain slopes that were once covered with majestic forests; industrial sewers are causing sterility in streams that used to teem with game fish; air pollutants generate opaque and irritating smogs that dull even the most brilliant and dramatic skies. The price of power, symbolized by superhighways and giant factories, is a desecration of nature and of human life.

Aggressive behavior for money or for prestige, the destruction of scenic beauty and historic landmarks, the waste of natural resources, the threats to health created by thoughtless technology—all these characteristics of our society contribute to the dehumanization of life. Society cannot be reformed by creating more wealth and power. Instead economic and technologic considerations must be made subservient to the needs, attributes, and aspirations that have been woven into the fabric of man's nature during his evolutionary and historical development.

The most hopeful sign for the future is the attempt by the rebellious young to reject our social values. Their protests indicate that mankind is becoming disturbed by increasing dehumanization and so may act in time to reverse the trend. Despite so many intellectual and ethical setbacks, despite so much evidence that human values are being spoiled or cheapened, despite the massive destruction of beauty and of natural resources, as long as there are rebels in our midst, there is reason to hope that our societies can be saved.

* * * *

Rebellion, however, should reach beyond conventional political and social issues. Even if perfect social justice and complete freedom from want were to prevail in a world at peace, rebels would still be needed wherever the world is out of joint, which now means everywhere. Rebellion permeates all aspects of human life. It originates from the subconscious will of mankind not to surrender to destructive forces. But rebelling is not the same as defining a cause that would improve the quality of human life, or formulating a constructive program of action. Marching in a parade is easier than blazing a trail through a forest or creating a new Jerusalem. . . .

Our society is highly expert in controlling the external world and even the human mind, but our relationships with other human beings and the rest of creation are constantly diminishing in significance. This society has more comfort, safety, and power than any before it, but the quality of life is cheapened by the physical and emotional junk heap we have created. We know that life is being damaged by the present social conditions, but we participate nevertheless in a system that spoils both the earth and human relationships. Most contemporary rebels, like the rest of us, are unwilling to give up the personal advantages so readily derived from the conditions we all know to be objectionable. Nevertheless, rebels play a useful social role; at least they voice our collective concern and make us aware of our collective guilt. But the acknowledgment of guilt is not enough.

Rumblings against the present state of things remain amorphous and ineffective largely because existing trends, customs, and policies cannot be changed merely by negative acts. Positive beliefs are required. Alternatives will not emerge through piecemeal evolution; their development demands an intellectual and emotional revolution. We cannot transform the world until we eliminate from our collective mind the concept that man's goals are the conquest of nature and the subjection of the human mind. Such a change in attitude will not be easy. The search for the mastery of nature and for unlimited growth generates a highly stimulating, almost intoxicating atmosphere, whereas the very hint of approaching stabilization creates apathy. For this reason, we can change our ways only if we adopt a new social ethic—almost a new social religion. Whatever form this religion takes, it will have to be based on harmony with nature as well as man, instead of the drive for mastery.

We have already accepted in principle, even though we rarely put into practice, the concept of human brotherhood. We must now take to heart the biblical teaching, "The Lord God took the man and put him into the Garden of Eden to dress it and to tend it" (Genesis 2:15). This means not only that the earth has been given to us for our enjoyment, but also that it has been entrusted to our care. Technicized societies thus far have exploited the earth; we must reverse this trend and learn to take care of it with love.

On the occasion of the annual meeting of the American Association for the Advancement of Science in 1966, the American historian Lynn White, Jr., pleaded for a new attitude toward man's nature and destiny. He saw as the only hope for the world's salvation the profoundly religious sense that the thirteenth-century Franciscans had for the spiritual and physical interdependence of all parts of nature. Scientists, and especially ecologists, he urged, should take as their patron Saint

Francis of Assisi (1182-1226).[1] But was not Francis one of the rebellious youths of his time—before the Church recognized that he was serving God by reidentifying man with nature? Francis, like Buddha, spent his early years in ease and luxury but rejected bourgeois comforts in search of more fundamental values. The contemporaries of both probably regarded them as beatniks.

The name Saint Francis and the word ecology are identified with an attitude toward science, technology, and life very different from that which identified man's future with his ability to dominate the cosmos. The creation of an environment in which scientific technology renders man completely independent of natural forces calls to mind a dismal future in which man will be served by robots and thereby himself become a robot. The humanness of life depends above all on the quality of man's relationships to the rest of creation—to the winds and the stars, to the flowers and the beasts, to smiling and weeping humanity.

Shortly before his death in 1963, the English novelist and essayist Aldous Huxley lamented on several occasions the fact that literature and the arts have not derived any worthwhile inspiration from modern science and technology. He thought the reason for this failure was that writers and artists are unaware of modern scientific and technological developments.[2] This may be part of the explanation but only a very small part. Like most other human beings, writers and artists are primarily concerned with perceptions, emotions, and values which the scientific enterprise must deliberately ignore. Yet scientists should not be satisfied with studying the biological machine whose body and mind can be altered and controlled by drugs and mechanical gadgets. They should become more vitally concerned about the nature and purpose of man. Only thus can they learn to speak to man not in a specialist's jargon but in a truly human language.

The New Pessimism

As the year 2000 approaches, an epidemic of sinister predictions is spreading all over the world, as happened among Christians during the period preceding the year 1000. Throughout the tenth century, Norsemen and Saracens incessantly raided Western Europe, disorganizing daily life and secular institutions, pillaging churches and monasteries. The rumor spread that the year 1000 would mark the end of the world and that a new spiritual universe would come into existence. Even those who did not believe that the world would come to an end probably assumed that living conditions would be corrupted by the barbaric invaders.

Prophets of gloom now predict that mankind is on a course of self-destruction, or that, in the unlikely event of its survival, it will progressively abandon the values and amenities of Western civilization. Nuclear warfare, environmental pollution, power blackouts, the progressive erosion of public services constitute direct and obvious threats to human existence. Furthermore, social regimentation and loss of privacy may soon reach levels incompatible with the traditional ways of civilized life. The established order of things appears to be threatened by technological and social forces that increasingly dominate the world, just as it was threatened by the raiding Norsemen and Saracens ten centuries ago.

Many observers of the contemporary scene would agree with the following words by the American journalist James Reston in the most influential daily newspaper of the most prosperous city in the world: "The old optimistic illusion that we can do anything we want is giving way to doubt, even to a new pessimism."[3] Newspaper headlines daily seem to confirm the belief that the problems of the cities, the races, and the nations are beyond our control.

Apprehension is most widespread and expresses itself most clearly with regard to nuclear warfare, threats to health, the rise of automation, and other ill-defined consequences of scientific technology. Popular articles entitled "The Truth About . . ." almost uniformly refer to the dangers of technological or medical innovations. The new pessimism, however, has other determinants which transcend the fear of annihilation and affect the quality of life. In particular, science is being accused of destroying religious and philosophic values without substituting other guides to behavior or providing a meaningful picture of the universe. The disintegrating effect of loss of belief was pungently expressed a generation ago by the American philosopher John Dewey in his warning that a culture which permits science to destroy traditional values, but which distrusts its power to create new ones, is destroying itself. Man finds it difficult to live without ultimate concern and faith in the significance of his destiny.

The malaise has now extended to the scientific community itself. While all scientists still believe that the opportunities for the extension of knowledge are boundless, many are beginning to doubt the wisdom and safety of extending much further some of the applications of knowledge.[4] In addition, there have been claims that limitations inherent in the very structure of the physical world may soon slow down, then interrupt altogether, the development of the scientific technologies which have resulted in the most spectacular achievements of our age. Airplanes cannot practically fly much faster than at the present supersonic speeds; electronic computers are approaching theoretical limits of speed and efficiency; high-energy accelerators cannot long continue to become larger and more powerful; even space travel will have achieved its human possibilities within a very few decades.[5]

The most important factor in dampening the euphoria that until recently was universal in scientific circles is the social and economic necessity of imposing directions and limitations to many technological developments. The current discussions concerning the advisability of devoting large resources to the manned space program have brought to light difficulties in reconciling the demands of certain technologies with more traditional human needs.

A few years ago, American scientists could state, "We *must* go to the moon, for the simple reason that we *can* do it"—echoing President John F. Kennedy, who in turn had echoed the statement by the English mountain climber George Mallory that Mount Everest *had* to be climbed, simply because it was there.[6] Such statements are admirable to the extent that they express man's determination to accept difficult challenges, whenever and wherever there is some chance that the effort will lead to spectacular feats. But dashing expressions do not constitute an adequate substitute for the responsibility of making value judgments.

There are many good scientific reasons for accepting the staggering human, financial, and technological effort required to explore space and to land a man on the moon. There are equally good reasons, however, for undertaking other kinds of difficult and challenging tasks—such as exploring the earth itself or the depths of the oceans, probing into the nature of matter and energy, searching for the origins of man and his civilizations, controlling organic and mental disease, striving for world peace, eliminating city slums, preventing further desecration of nature, or dedicating ourselves to works of beauty and to the establishment of an harmonious equilibrium between man and the rest of creation.

Laymen as well as scholars can think of many projects at least as important and interesting as space travel or lunar exploration, and just as likely to succeed. But limitations of resources make it impossible to prosecute all worthwhile projects at the same time. Hence, the statement that we *must* do something because we *can* do it is operationally and ethically meaningless; it is tantamount to an intellectual abdication. Like other responsible human beings, scientists and sociologists must discriminate; their choice of goals must be made on the basis of value judgments.

The problem of choice is greatly complicated by the fact that technological advances endlessly create new dilemmas, since every innovation has unforeseen consequences. Social regimentation, traffic jams, environmental pollution, constant exposure to noise and other unwanted stimuli are but a few of the undesirable accompaniments of economic and technological growth, Indeed, many innovations that have enhanced the wealth and power of our society in the past threaten to paralyze it at a later date. Abundance of goods, excess of comfort, multiplicity of means of communication are generating in the modern world situations almost as distressing as the ones that used to result from shortages of food, painful physical labor, and social isolation. We are creating new problems in the very process of solving those which plagued mankind in the past.

During recent years experts in the natural and social sciences have repeatedly pointed out that the erratic and misguided growth of technology and urban conditions now poses as serious a threat as the undisciplined growth of the world population. Economic affluence and scientific breakthroughs appear paradoxically to remove man still further from the golden age.[7]

The new pessimism derives in large part probably from the public's disenchantment at the realization that science cannot solve all human problems. Furthermore, the public is beginning to realize that whenever scientists make claims for support of their activities in the name of relevance to industrial technology, they are in fact making value judgments concerning the importance of technology in human life, judgments for which they have no special competence. A few spectacular technological failures might suffice to generate a bankruptcy of science.

Phrases such as the classical age, the age of faith, the age of reason, or the romantic age may not correspond to historical realities, but they convey nevertheless mankind's nostalgic longing for certain qualities of life that most people, rightly or wrongly, associate with the past. In contrast, we prosaically designate our own times the atomic age, space age, age of automation, antibiotic age—in other words, the age of one or another technology. These terms are used approvingly by technologists and disparagingly by humanists. The one term which has received almost universal acceptance is age of anxiety.

Social and technological achievements have spread economic affluence, increased comfort, accelerated transportation, and controlled certain forms of disease. But the material satisfactions thus made possible have not added much to happiness or to the significance of life. Not even the medical sciences have fulfilled their promises. While they have done much in the prevention and treatment of a few specific diseases, they have so far failed to increase true longevity or to create positive health. The age of affluence, technological marvels, and medical miracles is paradoxically the age of chronic ailments, of anxiety, and even of despair. Existentialist nausea has found its home in the most affluent and technologically advanced parts of the world.

Present-day societies abound in distressing problems, such as racial conflicts, economic poverty, emotional solitude, urban ugliness, injustice in all its forms, and the collective lunacy that creates the threat of nuclear warfare. But modern anxiety has deeper roots that reach into the very substance of each person's individuality. The most poignant problem of modern life is probably man's feeling that life has lost significance. The ancient religious and social creeds are being eroded by scientific knowledge and by the absurdity of world events. As a result, the expression "God is dead" is widely used in both theological and secular circles. Since the concept of God symbolized the totality of creation, man now remains without anchor. Those who affirm the death of God imply thereby the death of traditional man—whose life derived significance from his relation to the rest of the cosmos. The search for significance, the formulation of new meanings for the words God and Man, may be the most worthwhile pursuit in the age of anxiety and alienation.

Alienation is a vague word, but it denotes an attitude extremely widespread at present in affluent societies. Feeling alienated is an ancient experience which has taken different forms in the course of history. Many in the past experienced forlornness because the cosmos and the human condition appeared to them meaningless and pointless. Jean Jacques Rousseau, in the eighteenth century, traced alienation to the estrangement of man from nature that in his view resulted from artificial city life. Karl Marx, in the nineteenth, coined the word *Entfremdung*—rendered as "alienation" by his translators—to denote both the plight of the industrial worker deprived of the fruits of his production, and the depersonalization of labor in mechanized industries.

Many forms of alienation now coexist in our communities. The social and cultural malaise affects not only disenchanted intellectuals, industrial labor, and the poor classes, but also all those who feel depersonalized because circumstances compel them to accept mass standards which give them little chance to affirm their identity. Alienation is generated, furthermore, by the complete failure of even the most affluent societies to achieve harmonious relationships between human life and the total environment. The view that the modern world is absurd is no longer limited to the philosophical or literary avant-garde. It is spreading to all social and economic groups and affects all manifestations of life.

Psychologists, sociologists, and moralists tend to attribute anxiety and despair to the breakdown of intimate social relationships, with the attendant personal loneliness so pervasive in modern cities. The breakdown, however, is not limited to the interplay between human beings. It extends to the interplay between man and the natural forces that have shaped his physiological and mental self and to which his most fundamental processes are still bound. Chaos in human relationships has the same origin as chaos in the relationships between man and his environment.

In all countries of Western civilization, the largest part of life is now spent in an environment conditioned and often entirely created by technology. Thus one of the most significant and disturbing aspects of modern life is that man's contacts with the rest of creation are almost always distorted by artificial means, even though his senses and fundamental perceptions have remained the same since the Stone Age. Modern man is anxious, even during peace and in the midst of economic affluence, because the technological world that constitutes his immediate environment, by separating him from the natural world under which he evolved, fails to satisfy certain of his unchangeable needs. In many respects, modern man is like a wild animal spending its life in a zoo; like the animal, he is fed abundantly and protected from inclemencies but deprived of the natural stimuli essential for many functions of his body and mind. Man is alienated not only from other men, not only from nature, but more importantly from the deepest layers of his fundamental self.

The aspect of the new pessimism most commonly expressed is probably the belief that decrease in individual freedom is likely to result from increasing densities of population and the consequent need to accept a completely technicized urban environment. A heavy and repetitious anthology could be composed of writings by all kinds of scholars lamenting the sacrifice of personality and freedom at the altar of technological regimentation. As society becomes ever more highly organized, the individual will progressively vanish into the anonymous mass.

In his book *The Myth of the Machine*, the American critic Lewis Mumford predicts a future in which man will become passive and purposeless—a machine-conditioned animal designed and controlled for the benefit of depersonalized, collective organizations.[8]

* * * *

The new pessimism considers it almost inevitable that the complexity of social structures will result in social regimentation and that freedom and privacy will come to be regarded as antisocial luxuries. Under these conditions, the types of men more likely to prosper will be those willing to accept a sheltered but regimented way of life in a teeming and polluted environment from which both wilderness and fantasy will have disappeared. The world may escape catastrophic destruction, but if present trends continue our descendants will find it difficult to prevent a progressive decadence of the social order of things. The tide of events will bring about simultaneously, paradoxical as it may seem, the fragmentation of the person and the collectivization of the masses.

Naturally there are some optimists among the modern soothsayers, but the new Jerusalem they envision is little more than a dismal and grotesque magnification of the present state of affairs. They predict for America a gross national product of many trillions of dollars and an average family income so large that every home will be equipped with more and more power equipment and an endless variety of electronic gadgets. Drugs to control the operations of the body and the mind, complicated surgery, and organ transplantations will make it commonplace to convert ordinary citizens into "opti-men." The working day will be so short and the life span so long that countless hours will be available for the pursuit of entertainment—and eventually perhaps merely for the search for a *raison d'etre*.

Most modern prophets, in and out of the academies, seem to be satisfied with describing a world in which everything will move faster, grow larger, be mechanized, bacteriologically sterile, and emotionally safe. No hand will touch food in the automated kitchen; the care and behavior of children will be monitored electronically; there will be no need to call on one's friends because it will be possible to summon their voices, gestures, shapes, and complexions on the television-paneled walls of the living room; life will be effortless and without stress in air-conditioned houses romantically or excitingly lighted in all sorts of hues according to one's moods; exotic experiences will be safely and comfortably available in patios where artificial insect sounds and the proper degree of heat and moisture will create at will the atmosphere of a tropical night or a New England summer day. Actual examples of the dismal life that technological prophets envision for the future can be found increasingly in periodicals and books, notably the recently published *The Year 2000*. . . .[9]*

* * * *

The word "unbelievable" (or its equivalent "incredible") is ambiguous. As commonly used, it denotes events or situations so extraordinary that they are difficult to believe but are nevertheless true. The present writings about all the marvels of the "unbelievable future" intend to convey this sense of impossible but true. Etymologically, however, unbelievable has a much more negative meaning, and it is in the sense of actual impossibility that the word will be used here. The "opti-man" imagined by the prophets of dismal optimism turns out to

*Some of the footnotes have been deleted along with portions of the original text. For the purpose of this reproduction the footnotes have been renumbered to read consecutively.

be not only a hollow man, but also a pseudo-man; not only would he be devoid of the attributes that have given its unique value to the human condition, but he would not long survive because he would be deprived of the stimuli required for physical and mental sanity.

The kind of life so widely predicted for the twenty-first century is unbelievable in the etymological sense because it is incompatible with the fundamental needs of man's nature. These needs have not changed significantly since the Late Stone Age and they will not change in the predictable future; they define the limits beyond which any prediction of the future becomes literally unbelievable.

Whatever scientific technology may create, *l'homme moyen sensuel* will continue to live by his senses and to perceive the world through them. As a result, he will eventually reject excessive abstraction and mechanization in order to reestablish direct contact with the natural forces from which he derives the awareness of his own existence and to which he owes his very sense of being.

The one possible aspect of the future seldom discussed by those who try to imagine the world-to-be is that human beings will become bored with automated kitchens, high-speed travel, and the monitoring of human contacts through electronic gadgets. People of the year 2000 might make nonsense of the predictions now being published in the proceedings of learned academies and in better-life magazines, simply by deciding that they want to regain direct contact with the natural forces that have shaped man's biological and mental being. The visceral determinants of life are so permanent, and so demanding, that mankind cannot long safely ignore them. In my opinion, the world in the year 2000 will reflect less the projections of technologists, sociologists, and economists than the vital needs and urges of biological man.

At the end of his *Education*, written in 1905, the American historian Henry Adams gloomily predicted that the cult of the Dynamo was to be the modern substitute for the cult of the Virgin.[10] The present scene appears to confirm his prediction, but the future may still prove him wrong. One begins to perceive disenchantment among the worshipers of the Dynamo, and, more importantly, there are encouraging signs of unrest in the younger generation. Frequently in the past the son rejected what the father had taken for granted, and civilization thus took a step forward. Beatniks, hipsters, teddy boys, provos, hooligans, *blousons noirs*, and the countless other types of rebellious youths are probably as ignorant, foolish, and irresponsible as conventional people believe them to be. But conventionality rarely has the knack of guessing who will shape the future. The substantial citizen of Imperial Rome and the orthodox Jews of the synagogue looked down on the small tradesmen, fishermen, beggars, and the prostitutes who followed Jesus as he preached contempt for the existing order of things. Yet Imperial Rome and the Temple collapsed, while Jesus' followers changed the course of history.

The vision of the future, as seen in the light of the new intellectual pessimism or of the dismal optimism of some technologists, would be terribly depressing if it were not for the fact that it so much resembles visions of the future throughout history. Pessimists have repeatedly predicted the end of the world, and utopians have tried to force mankind into many forms of straitjackets. Those who did not live before 1789, wrote the French statesman Talleyrand, have not known the *douceur de vivre*. This melancholy belief did not prevent Talleyrand from living very successfully to the age of eighty-four. When he died in 1838, the storming of the Bastille was half a century past. The Industrial Revolution had begun, and the world certainly looked uncouth and dark to many genteel souls. But we now realize that it was a beginning rather than an end. Like Talleyrand, and like society after the Industrial Revolution, we too shall proably manage to find an acceptable formula for our times. Fortunately, the creativeness of life always transcends the imaginings of scholars, technologists, and science-fiction writers.

Toward a New Optimism

Despite the forebodings of the tenth century, the world did not come to an end in the year 1000, nor did Europe take to barbaric ways of life. The Saracens assimilated Greek learning and transmitted it to the Mediterranean universities, from which it spread all over the Western world. The Norsemen became Christianized, and, far from destroying civilization, their uncouth barons created monasteries, churches, cathedrals, and town halls almost as fast as they built fortified castles. The Norman rulers spread first Romanesque and then Gothic architecture all over Europe to honor the Virgin and the saints; in many places also their courts provided a chivalrous atmosphere in which the troubadours converted the worship of the Virgin into the cult of womanhood.

If the rebellious young succeed in discovering a formula of life as attractive as that of the troubadours, we may witness in the twenty-first century a new departure in civilization as occurred in Europe after it recovered from the fears of the tenth century. To be humanly successful, the new ages will have to overcome the present intoxication with the use of power for the conquest of the cosmos, and to rise above the simple-minded and degrading concept of man as a machine. The first move toward a richer and more human philosophy of life should be to rediscover man's partnership with nature.

The undisciplined and incoherent expansion experienced by technicized societies during the past few decades would certainly spell the end of the human condition if it were to continue much longer. Doing more and more of the same, at a faster and faster pace, contributes neither to happiness nor to the significance of life. In the past, great prosperity has often damaged human values and generated boredom; the environmental crisis in the modern world indicates, furthermore, that mismanaged prosperity may destroy human life altogether.

The fact that economic affluence commonly leads to absurdity had been noted in the nineteenth century by Ralph Waldo Emerson, who predicted in his journal that American prosperity "would go on to madness." Many now believe that modern life in large cities is dangerously close to the state of madness!

Emerson was far-sighted when he wrote these words, because until the 1940s political reformers, economists, technologists, and scientists had no reason to question the usefulness of their efforts. Social action based on objective knowledge then appeared unequivocally beneficial; it commonly resulted in greater personal freedom, made life safer and more comfortable, protected man from irrational fears by enlarging his view of the cosmos and his understanding of biological processes. Until our times, moving onward with scientific technology could be legitimately identified not only with the creation of wealth, but also with social progress.

North America provided an ideal setting for the euphoria of the nineteenth and early twentieth centuries, which was based on the belief that industrial civilization would inevitably generate happiness by increasing comfort and creating more, better, and cheaper goods. The vastness and emptiness of the continent made it easy for the settlers to accept the myth of the ever-expanding frontier. The mood of optimistic nomadism which has been so influential in shaping American attitudes and institutions certainly derived in large part from the nineteenth-century faith that one could always move on to greener pastures. After the whole continent had been occupied, the explosive development of science and technology provided grounds for even greater optimism by opening exciting vistas for knowledge and for technological enterprise and thus providing new, apparently endless frontiers for economic growth.

* * * *

Since 1950, the urge for economic growth has been increasingly overshadowed by public concern with the undesirable consequences of growth: crowding, environmental pollution, traffic jams, surfeit of goods, and all the other nauseating and catastrophic by-products of excessive population, production, and consumption. Men of the twentieth century may still be whistling on their way, but deep in their hearts they are worrying about where they should go. Often they are not even sure whether they should keep on going or try to retrace their steps.

There are several reasons for the widespread skepticism concerning the advantages and even the possibility of unlimited technological growth. One is the awareness already mentioned that beyond a certain point prosperity and abundance of goods become meaningless. It is increasingly apparent, furthermore, that certain present trends are self-limiting because they lead to absurdities which, if continued, generate countertrends. The growing interest in crafts, home cooking, folk dances, and the various forms of "be-ins" certainly represents a trend against the standardization of industrial goods and commercial entertainment. The flow of population from the heart of the city to suburbia, then to exurbia, and then back to the city may be another example of trend-countertrend.

The view that Western civilization must abandon its growth myth should not be confused with the thesis of the German philosopher Oswald Spengler in *The Decline of the West* (1918) that the Western world cannot escape decadence. Rather it constitutes an expression of my faith that Western culture, and especially Western science, can be rededicated to values more lasting and more significant than those heretofore identified with technological and economic growth. The new optimism finds its sustenance in the belief that science, technology, and social organization can be made to serve the fundamental needs and urges of mankind, instead of being allowed to distort human life.

There is fear of science among the general public and resentment against it among classical scholars. But stronger and more widespread than this hostility is the belief that scientific techniques will be needed to solve the world's problems, including those created by scientific technology itself. Witness the insistent demands from the executive branch of the government, from Congressional committees, and from various private organizations that scientists direct their efforts more pointedly to the problems of man in the modern world. . . .

Ever since the seventeenth century, science has been concerned primarily with atomistic descriptions of substances and phenomena. Its philosophical heroes have been Democritus (fifth century B.C.) and René Descartes (1596-1650), both of whom taught that the way to knowledge is to separate substances and events into their ultimate components and reactions. The most pressing problems of humanity, however, involve relationships, communications, changes of trends—in other words, situations in which systems must be studied as a whole in all the complexity of their interactions. This is particularly true of human life. When life is considered only in its specialized functions, the outcome is a world emptied of meaning. To be fully relevant to life, science must deal with the responses of the total organism to the total environment. An earlier Greek philosopher, Heraclitus, who taught that everything is flux, may well replace Democritus as the precursor of the new scientific humanism.

We are worried by the universal threat to natural resources and shudder at the raping of nature caused by scientific technology and overpopulation. We wonder, indeed, whether man can long survive in the artificial environment he is creating. To approach these problems constructively, we must learn more of the complex interplays between man, his technologies, and his environment. We must define with greater precision the determinants of man's responses to environmental forces—his innate limitations as well as his potentialities, his acquired characteristics as well as his aspirations. A sophisticated form of ecology will have to complement Democritus' atomicism and Descartes' reductionism.

We lament the dehumanization of man. Anthropology has taught us that man acquired his humanness while evolving in intimate relation with other living things and we know that all phases of his development are still conditioned by the social stimuli that he receives in the course of his life. We must develop a science of modern man considered not as an object, but rather in his interplay with other human beings—during both the emotional depth of individual encounters and the less demanding ordinary social relationships.

Since science and the technologies derived from it are now playing such an immense role in human societies and changing them so rapidly, man can survive only by continuously adjusting himself to ever-new conditions. Such adaptive processes are the inevitable consequences of social and technological innovations. Many modern thinkers—biologists as well as technologists, religious believers as well as atheists—go so far as to state that man's increasing dependence on the machine constitutes an essential fact in his evolution—the process that the Jesuit archaeologist Pierre Teilhard de Chardin has termed hominization.[11] Despite the authority this concept has thus received from theologians, philosophers, scientists, and enlightened laymen, the view that man's future is linked to technology can become dangerous if accepted uncritically. Any discussion of the future must take into account the inexorable biological limitations of *Homo sapiens*.

Acknowledgment of these limitations need not imply either a static view of man's nature or a resigned acceptance of the status quo for the human condition. Looking by night on the towering black mass of Chicago's buildings, the American architect Frank Lloyd Wright came as early as 1901 to the conclusion that "if this power must be uprooted that civilization may live, then civilization is already doomed."[12] Scientific technology cannot and should not be uprooted; not only has it become indispensable for man's survival but it has enriched his perceptions, enlarged his vision, and deepened his concept of reality. To a very large extent the continued unfolding of civilization will depend on the imaginative creativity of scientific technologists. But it would be dangerous to assume that mankind can safely adjust to all forms of technological development. In the final analysis, the frontiers of social and technological innovations will be determined not by the extent to which man can manipulate the external world but by the limitations of his own biological and emotional nature.

Total wisdom requires the attitudes of both the sage and the scientist, integrated on the high ground of man's meta-technical being and destination. Pure rationalism degrades wisdom to the level of bloodless abstractions, and technocratic thinking reduces man to a machine. On the other hand, science in all its forms and applications is now creating values that transcend the mechanical aspects of life. A social philosophy suitable for our times must therefore include scientific humanism. Our formidable knowledge of physical forces and inanimate matter must be supplemented by scientific knowledge of the living experience and by awareness of human aspirations.

In his Sigma Xi—Phi Beta Kappa address before the annual meeting of the American Association for the Advancement of Science in 1966, the American astronomer W. O. Roberts dared to raise questions about the nature and purpose of man and about what constitutes a good life and a good society. A generation ago such questions would have been considered outside the province of scientists. But Dr. Roberts was expressing the universal uneasiness about the future when he pleaded for a concern with ultimate purpose in human life and for a philosophy geared to the chain-reacting growth of science. He referred to science as "a wellspring of our discontent," not because of its obvious influence on the practical aspects of our day-to-day life, but because of its impact on man's changing conception of himself and his world.[13]

In addition to the science of material things we must develop a science of humanity. Both together will constitute the humanism of the future, a new kind of *Gai Savoir.*[14]

Reference Notes

1. Lynn White, Jr., "The Historical Roots of Our Ecologic Crisis," *Science*, 155 (1967), 1203-1207.

2. Aldous Huxley, *Literature and Science* (New York: Harper, 1963), and "Achieving a Perspective on the Technological Order," in Carl F. Stover (ed.), *The Technological Order* (Detroit, Mich.: Wayne State University Press, 1963, 252-258).

3. James Reston, "Washington: The New Pessimism," *The New York Times*, April 21, 1967, 38; see also editorial, "Voices of Doubt," *The Wall Street Journal*, April 26, 1967, 16.

4. There have been many expressions of this attitude in recent scientific and sociological literature; see Stover, *op. cit.;* Jacques Ellul, *The Technological Society*, trans. John Wilkinson (New York: Knopf, 1965); Elmer Engstrom, "Science, Technology, and Statesmanship," *American Scientist*, 55 (1967), 72-79; White, *op. cit.;* also the discussion "Does Science Neglect Society?", *Science*, 158 (1967), 1134-1136.

5. John R. Platt, *The Step to Man* (New York: Wiley, 1966), 185-203.

6. Vernon Van Dyke, *Pride and Power: The Rationale of the Space Program* (Urbana, Ill.: University of Illinois Press, 1964), 155.

7. See Stover, *op. cit.;* Ellul, *op. cit.*

8. Lewis Mumford, *The Myth of the Machine* (New York: Harcourt, 1967).

9. Herman Kahn and Anthony Wiener, *The Year 2000, a Framework for Speculation on the Next Thirty-three Years* (New York: Macmillan, 1968).

10. Henry Adams, *The Education of Henry Adams* (Boston: Houghton Mifflin, 1906).

11. Pierre Teilhard de Chardin, *The Phenomenon of Man*, trans. Bernard Wall (New York: Harper, 1959); French title: *Le Phénomène Humain.*

12. Frank Lloyd Wright, *The Future of Architecture* (New York: Horizon Press, 1953), 80.

13. W. O. Roberts, "Science, A Wellspring of Our Discontent," *American Scientist*, 55 (1967), 3-14.

14. I have benefited in writing this chapter from discussions with Mrs. Julie Field, author, with Will Burtin, of a manuscript on "The Architecture of an Ethic."

The Environmental Crisis

Neil H. Jacoby

Who would have predicted, even as recently as a year ago, the strong ground swell of public concern about the environment that now preoccupies Americans? The great silent majority as well as activists of the left have discovered that our country is running out of clean air and pure water. Suddenly, we all understand that smog, noise, congestion, highway carnage, oil-stained beaches, junk graveyards, ugliness, and blatant commercial advertising not only offend our senses but threaten our health and our very lives.

Now we are trying to identify the culpable parties and to demand corrective action. What are the basic forces behind environmental deterioration and why has a crisis emerged so swiftly? What are the merits of the diagnoses and prescriptions that have been advanced for the environmental problem? How can the environment be improved, and who should pay the costs? What are the respective roles and responsibilities of business and of government in restoring environmental amenities? Above all, what lessons does the environmental crisis teach about the functioning of our political and market systems, and about reforms needed to forestall other crises in the future?

We focus attention upon the urban physical environment, that is, upon the spatial and sensory qualities of the land, air, water, and physical facilities that surround the three out of four Americans who live in towns and cities. This milieu deteriorates as a result of air and water pollution, noise, industrial and household waste materials, declining quantity or quality of housing per capita, crowding, congestion, loss

Reprinted from *The Center Magazine* (November/December, 1970), pp. 36-48. Chart on p. 309 has been redrawn.

of privacy and recreational facilities, rising accidents and loss of time in urban transportation, and, not least of all, drabness and ugliness.

The physical environment is, of course, only one dimension of the quality of human life. In focusing upon physical factors, one excludes important social and psychological factors such as order and security, social mobility, and the social participation or alienation of the individual. All of these environmental factors, along with per capita income, wealth, health, and education, need enhancement.

Spatially, the urban environment must be viewed as one subdivision of the entire global ecosystem, which also embraces rural lands, the oceans, the atmosphere surrounding the earth, and outer space. Since all parts of this system interact, ideally it should be analyzed, planned, and managed as a whole.

The urban physical environment nevertheless merits a top priority because it affects the majority of our population and, by general assent, its qualities are below the threshold of tolerability. In addition, physical factors powerfully influence the health, mental attitudes, and life-styles of urban residents, and their enhancement will elevate the social and psychological qualities of American society. One is therefore justified in focusing attention upon the physical characteristics of urban life, notwithstanding that it is a partial analysis of the global ecosystem.

Three basic forces have operated to change the urban physical environment for the worse: population concentration, rising affluence, and technological change. The overwhelming tendency of people to concentrate in cities has worsened the environment in many ways. Traffic congestion, crowding, overloading of transportation, marketing and living facilities, delays and loss of time, along with rising levels of air, water and noise pollution, have been among the social costs of urbanization. During the half-century between 1910 and 1960 the percentage of Americans living in urban areas of 2,500 or more rose from 45.7 to 70, while the number of urbanites tripled from 42 to 125 million. Beyond doubt, the 1970 census will reveal an accelerated urbanization. Urbanization clearly brings benefits to people—wider job opportunities, richer educational and cultural fare, more individual freedom from social constraints—or else it would not have been so powerful and enduring a movement. Yet, beyond some levels of population size and density, the total costs of urbanization begin to exceed the total benefits. Discovery of the optimum size of cities and optimum density of their populations are vitally important tasks confronting national planners.

A second prime mover in environmental change has been rising affluence—the expansion of annual real income and expenditure per capita. Real income per person (measured in 1958 dollars) more than doubled during the eighteen years, 1950-1968, from $1,501 to $3,409. As real incomes have mounted, each person has bought and consumed more tangible goods, thrown them away more quickly, and generated solid waste. Each person has traveled more miles per year, multiplied his contacts with other people, and rapidly expanded his usage of energy. All of this has increased air, water, and noise pollution, crowding, and congestion, traffic accidents. With the number of urbanites doubling and per capita real incomes quadrupling every forty years, the problem of supplying urban amenities is exploding. One shudders to contemplate the environmental degradation that would occur if 525 million Indians, now crowded 417 per square mile, were each to spend as much as 200 million Americans living only 60 per square mile. India seeks affluence, but could she stand it?

Environmental degradation is not, of course, inherent in rising affluence. Only the particular forms and methods of production and consumption to which our society has become accustomed degrade it. Rising affluence can and should be a source of environmental enhancement.

It is often overlooked that rising per capita income results in an increased demand for environmental amenities. People naturally demand better public goods—more comfort and convenience and beauty in their communities—to match the better private goods and services their rising incomes enable them to buy. One reason for the environmental "crisis" is the frustration felt by the public with a short supply of environmental amenities available to meet a rising demand for them.

The physical environment of large American cities has not degenerated absolutely in an overall sense, but probably has been improving. People easily forget amenities taken for granted today that were lacking half a century ago. Examples are air-conditioned offices, restaurants, and homes; thermostatically controlled electric and gas heat; underground utility wires and poles; paved boulevards and auto freeways. These have widely replaced the crowded slums, the filth of unpaved streets, the drafty cold-water flats and belching chimneys of winter, and the steaming miseries of unrefrigerated summers. Even in the inner city, people today live longer, healthier, and more comfortable lives—if not happier ones—than they did before World War I. What has happened is that the overall supply of urban amenities has fallen far short of the rising effective demand for them, and the supply of certain critical goods, such as pure air and water, has virtually vanished.

The third source of the environmental problem is technological change. Advancing technology has expanded the variety of products available for consumption, made products more complex, raised rates of obsolescence and thereby added to waste disposal. It has also added immensely to the per capita consumption of physical materials and energy, with consequent increments of waste and of pollution. It has expanded the amount of information required by consumers to make rational choices in markets thereby creating market imperfections that are the source of the contemporary "consumerism" movement. Technological change, however, is like rising affluence, a two-edged sword; it can be used to improve as well as to degrade the environment. Technology can *reduce* material consumption and recycle harmful wastes. Examples are the replacement of bulky vacuum tubes by microminiaturized circuits in computers, or the conversion of sewage into pure water plus fertilizers. Environmental preservation calls for a redirection of our technological efforts, as well as a restructuring of patterns of consumption.

One conspicuous aspect of environmental deterioration has been the disappearance of "free goods"—amenities such as clean air, pure water, and open space—that are in such ample supply relative to the demand for them that they are not economized. Pure air is no longer free. To obtain it one must buy air-conditioning equipment and acquire a home in which to install it. Pure water must be purchased by the bottle, now that the product of many municipal water systems is barely potable. Most urban dwellers must spend large sums of money for travel in order to gain the privacy and recreation of a natural environment unavailable at home.

A second aspect of environmental change is the fast-rising importance of spatial relationships in the cities. Such factors as building heights and population densities, street layout, park location, and zoning patterns largely determine the life-styles of urban residents and the supply of amenities available to them. The atrociously bad planning of most American cities and the abject perversion of zoning and building requirements to serve short-term commercial interests are well documented. The flagrantly over-dense building on Manhattan Island has been permitted only because of popular ignorance and apathy. Now, the public is belatedly recognizing the heavy social costs that its neglect has created. Popular concern with city planning, zoning, and building development is rising. The heavy stake of the individual in the physical attributes of his community is finally being appreciated.

A third aspect of environmental change is the multiplication of interdependencies among individuals. To an increasing extent the activities of each of us impinge upon others. This is so not only because more people live in cities, but also because the scale and variety of each person's activities rise with the amount of real income he produces and consumes. Thus, no one suffered disamenity a generation ago when his neighbor played a phonograph in a suburban home; but many suffer when a neighbor's son now turns up the sound volume of his hi-fi instrument in a high-rise apartment building.

Increasing interdependency is one way of looking at what economists call the "spillover effects" or external costs of production or consumption. For example, paper mills emit chemical wastes into lakes and streams, copper smelters inject sulphur dioxide into the air, and electric generating stations throw off carbon monoxide, radioactive wastes, or hot water, depending upon their fuels. Motor vehicles cause massive air and noise pollution, traffic accidents, and vast expenditures on medical, legal, policing, and engineering services and facilities—all borne mainly by the public. These industries all generate external costs, thrust upon society in the form of loss of environmental amenities. Although reliable estimates are lacking, total external costs in the U.S. economy are of the order of tens of billions of dollars a year.

The speed with which public interest in the environment has mounted may be explained primarily by the swift decline in certain amenities below thresholds of tolerability. Although certain critical amenities, notably pure air, have been diminishing for many years, the public has suddenly become aware of critical deficiencies. Thus, the quality of air in the Los Angeles Basin deteriorated steadily after 1940. Yet only by the mid-nineteen-sixties, after school children were being advised not to exercise outdoors on smoggy days and when smog alerts were being sounded on many days each year, was decisive action taken to reduce air pollution from motor vehicles. By the sixties, people saw that the "capacity" of the atmosphere over the basin to disperse pollutants had been intolerably overloaded.

After the design capacity of any facility has been reached, amenities diminish exponentially with arithmetic increases in the load. For example, when a twenty-first person enters an elevator designed to hold twenty persons, everyone in the elevator suffers loss of comfort; and when a twenty-second person enters, the percentage loss of amenity is much greater than the 4.8 per cent increase in the number of passengers. Similarly, when the five thousand and first automobile enters a freeway designed to carry five thousand vehicles per hour, it puts pressure of inadequate space upon five thousand and one drivers, and not only upon the new entrant.

Another reason for current public concern with the environment is the gathering appreciation of inequity as some groups in society gain benefits at the cost of other groups. The automobilist whose vehicle spews out air pollution gets the benefits of rapid and convenient travel; but he imposes part of the costs of that travel upon people who are forced to breathe bad air and hear deafening noise and who must bear the costs of painting and maintaining property corroded by pollutants. Because this is manifestly inequitable, upgrading the environment by eliminating this kind of pollution will not only add to aggregate real income, but will also improve its distribution.

Before examining effective measures for enhancing the environment let us dispose of a number of partial or superficial diagnoses of, and prescriptions for, the problem. Several schools of thought have arisen.

First, there is the Doomsday School. It holds in effect that the problem of environmental degradation is insoluble. For example, Paul Ehrlich argues in his book *The Population Bomb* that it is already too late to arrest man's inexorable march to racial extinction through overpopulation, malnutrition, famine, and disease. Other criers of doom are the natural scientists who predict changes in the earth's temperature, as a result of accumulating carbon dioxide in the atmosphere, with consequent melting of the polar ice and other disasters. Although laymen are incompetent to judge such matters, they remain moot issues among natural scientists and therefore call for at least suspended judgment. Accumulating evidence suggests that population growth in the advanced nations has already slowed appreciably, and is starting to do so in many less developed lands. In any event, an apocalyptic view of the future should be rejected if only because it leads to despair and inaction. If one really believes that the future is hopeless, one will cease making an effort to improve society.

At the opposite pole is the Minimalist School. It holds that environmental deterioration is a minor problem in comparison with such contemporary issues as poverty, civil rights, and school integration. Its members argue that political leaders calling for a better environment are "eco-escapists," seeking to divert public attention from their failure to resolve these primary issues. What the Minimalists overlook is that the United States is already making progress in reducing poverty, expanding civil rights, and achieving educational integration, while it is still losing ground in arresting the decline of the urban environment. They also forget that attention to the environment does not mean neglect of poverty. On the contrary, central-city areas generally have the worst physical conditions of life and are populated mainly by low-income families. Because the poor stand to gain most from environmental enhancement, a war on pollution is one battlefront in a war on poverty. A vigorous attack on that front need not inhibit action on other fronts.

There is also a Socialist School. Its members view environmental deterioration as an inescapable consequence of capitalist "exploitation." If only private enterprise, market competition, and profit incentives were replaced by central planning and state ownership and management of enterprises, they contend, the problem would disappear. However, the socialist countries are facing more serious problems of pollution as their per capita G.N.P.'s are rising. Managers of socialist enterprises are judged by the central planners on the efficiency of their operations, and are under as much pressure to minimize internal costs and throw as much external cost as possible on the public as are the managers of private firms in market economies who seek to maximize stockholders' profits. Moreover, because a monolithic socialist society lacks a separate and independent mechanism of political control of economic processes, it is less likely to internalize the full costs of production than a market economy, with its dual systems of market-price and governmental controls. Pollution has arisen primarily from the failure of our political system, acting through government, to establish desired standards of production and consumption. If government performs its unique tasks, the competitive market system will operate within that framework to produce what the public demands without harming the environment.

The largest group of new environmentalists appear to be associated with the Zero Growth School. Its thesis is simple: since environmental degradation is caused by more people consuming more goods, the answer is to stop the growth of population and production. Nature has fixed the dimensions of the natural environment; therefore man should fix his numbers and their economic activities. We must establish a stable relationship between human society and the natural world.

Zero economic and population growth could arrest the process of environmental degradation, but could not, per se, restore a good physical environment. Were real G.N.P. constant through time, current levels of air and water pollution, noise, crowding, ugliness, and other negative elements would continue as long as present patterns of production and consumption are maintained.

Zero growth of population and production is, moreover, impossible to achieve. Because economic growth is a product of expanding population, higher investment, and advancing technology, zero growth would call for stopping changes in all three variables. This cannot be done in the proximate future, if at all. A leading population analyst has shown that even if, beginning in 1975, every family in the United States were limited to two children—an heroic assumption—population dynamics are such that this nation would not stop adding people until about 2050 A.D., when it would contain nearly 300 millions. (See Stephen Enke, "Zero Population Growth—When, How, and Why," TEMPO Publication 70TMP35, Santa Barbara, California, June 2, 1970.) While a decline in net savings and investment to zero is possible, it is extremely unlikely in view of the savings and investment rates Americans have maintained during the present century in the face of enormous increases in their real wealth and incomes. (See *Policies for Economic Growth and Progress in the Seventies,* Report of the President's Task Force on Economic Growth, U.S. Government Printing Office, Washington, D.C., 1970.) A static technology of production is inconceivable. As long as Americans remain thinking animals they will increase the productivity of work.

Finally, zero growth is undesirable. A rising G.N.P. will enable the nation more easily to bear the costs of eliminating pollution. Because zero growth of population is far in the distance, and zero growth of output is both undesirable and unattainable, it follows that the environmental problem must be solved, as President Nixon stated in his January, 1970, State of the Union Message, by redirecting the growth that will inevitably take place.

The Austerity School of environmental thought is related to the Zero Growth School. Its members assert that environmental decline is produced by excessive use of resources. They are outraged by the fact that the United States consumes about forty per cent of the world's energy and materials, although it contains only six per cent of the world's population. Believing that asceticism is the remedy, they call for less consumption in order to conserve resources and to reduce production and pollution. We should convert ourselves from a society of "waste-makers" into one of "string-savers."

The basic error here is that it is not the amount of production and consumption per capita that degrades the environment, but the fact that government has failed to control the processes of production and consumption so as to eliminate the pollution associated with them. Without such political action, consumption could be cut in half and society would still suffer half as much pollution; with appropriate political control consumption could be doubled while pollution is radically reduced. The second error of the Austerity School, which distinguishes it from the Zero Growth School, is a notion that the world confronts a severe shortage of basic natural resources. Exhaustive studies by Resources for the Future have shown the contrary: there are no foreseeable limitations upon supplies of basic natural resources, including energy, at approximately current levels of cost. Technological progress is continually opening up new supplies of materials that are substitutable for conventional materials (e.g., synthetic rubber and fibers) and lowering the costs of alternative sources of energy (e.g., production of petroleum products from oil shales, tar sands, and coal). Austerity theorists do make a valid point, however, when they observe that governmental regulation to internalize external costs can cause business enterprises to develop ways of recycling former waste materials back into useful channels.

Finally, there is the Public Priorities School. Its adherents see the problem as one of too much governmental spending on defense and space exploration, leaving too little for environmental protection. The solution, as they see it, is to reallocate public expenditures. There are two responses to this line of reasoning: public expenditures are already being strongly reordered, and in any event reallocations of private expenditures will weigh far more heavily in a solution of the environmental problem. Thus between the fiscal years 1969 and 1971 federal budget outlays on defense and space are scheduled to shrink by ten per cent, from $85.5 billions to $77 billions, whereas outlays on social security and public assistance will rise by twenty-six per cent, from $46 billions to $60 billions. The President has announced plans for further contractions of defense outlays and expansions of expenditures on the nation's human resources.

Environmental restoration does require large increases in public expenditures upon sewage disposal and water purification, parks, housing, urban development, and public transportation. Even more, however, it calls for a reallocation of private expenditures as a result of governmental actions to internalize external costs in the private sector. For example, the purchase price and operating expenses of an automobile that is pollution-free will undoubtedly be higher than for a vehicle that degrades the environment, because the auto user will be paying the full costs of his private transportation. With internalization of costs, spending on private auto transportation may be expected to decline relatively. At the same time, spending on education and housing, which produce external benefits, will increase relatively. In the aggregate, readjustments in patterns of private expenditure will far outweigh reallocation of public expenditure in a total program of environmental restoration.

Because the environmental problem is critically important and is soluble, and neither socialization of the economy, zero growth, austerity, nor new public spending priorities offer a satisfactory solution, a more basic approach must be made. A good policy for environmental improvement should improve the distribution of income among people as well as the allocation of society's resources. Governmental intervention is necessary to attain both ends.

Environmental degradation occurs, as has been shown, when there are significant external costs involved in producing or consuming commodities. A social optimum cannot be achieved when there is a divergence between private (internal) and social (external plus internal) costs. An optimal allocation of society's resources requires that the full costs of production of each good or service be taken into account.

The internalization of external costs must therefore be a pivotal aim of environmental policy. (A trenchant description of the external costs of economic growth is given by E. J. Mishan, *The Costs of Economic Growth*, New York: Praeger, 1967.)

Theoretically, perfectly competitive markets in which there are no transaction costs will lead to an optimum reallocation of resources in cases of pollution via bargaining between the polluter and the person harmed by pollution, no matter which party is legally responsible to compensate the other. (See R. H. Coase, "The Problem of Social Costs," *Journal of Law and Economics*, III, October, 1960.) In practice, however, the transaction costs of education, organization, and litigation are excessively high when pollution affects large numbers of people, as it usually does. For this reason it is more efficient for government to resolve pollution problems by legislation or regulation, rather than to leave them to bilateral market bargaining. For example, government can order air polluters to reduce their emissions by x per cent. Polluters then incur (internalize) costs in order to conform to the public regulation, thereby relieving the public of even greater costs of maintaining health and property damaged by pollution.

Prior governmental action is essential because the competitive market system is incapable, by itself, of internalizing the costs of antipollution measures. Suppose, for example, that the automobile could be made pollution-free by installing a device costing x dollars. An automobile owner would not voluntarily install the device, because other people would reap the benefits of the cleaner air made possible by his expenditure. General Motors proved this in 1970 by a well-advertised effort to sell motorists in the Phoenix, Arizona, area a pollution-reducing kit costing only twenty-six dollars. During the first month only a few hundred kits were sold in a market with several hundred thousand potential buyers. Auto makers would not voluntarily install the device because to do so would add to their costs and put them at a disadvantage in competition with other manufacturers who did not install it. And antitrust laws prohibit any agreement among all auto manufacturers simultaneously to install, or not to install, pollution-reducing devices. Where large external costs or benefits are involved, there is a conflict between the decision that serves the self-interest of the individual and that which serves the collective welfare of the community. Community welfare can only be given the precedence it deserves by a prior governmental action regulating private behavior, followed by corporation actions to modify products, prices, and allocations of resources in order to conform to the public regulation.

Society cannot reasonably expect individual enterprises or consumers to shoulder external costs in the name of "social responsibility," because the competitive market system puts each firm and household under strong pressure to minimize its costs in order to survive. What is needed is a prior political decision that leaves all producers or consumers in the same relative position.

There are usually alternative solutions to pollution problems; each alternative should be evaluated in order to identify the least costly of them. Consider again the example of smog in the Los Angeles Basin. Among possible ways of coping with this problem are the following: controlling emissions of pollutants from motor vehicles and stationary sources by public regulation; moving people out of the basin; rezoning to reduce building density; building a rapid mass-transit system; imposing heavy taxes on private automobile operation; or subsidizing motorists to limit their auto mileage. The costs and benefits of each alternative, and combinations thereof, should be evaluated before an anti-pollution policy is adopted. The goal should always be the most efficient use of scarce resources.

All desirable things in limited supply have a cost, and there are trade-offs between desirable things. People may gain more of one thing only by sacrificing something else, and the optimum situation is reached when no additional benefits can be obtained by further substitutions. These principles apply to environmental amenities. For example, noise pollution can be reduced with benefits to health and well-being, but at the cost of larger expenditures for insulation or noise-abatement devices or a reduction in the speed or power of engines. Conceivably, utter silence could be achieved by incurring astronomical costs and by making great sacrifices of mobility, power, and time. The public decides the optimum noise level by balancing the benefits of less noise against the costs of attaining it. Government then fixes a noise standard at that point where the costs of reducing noise further would exceed the additional benefits to health and well-being. Although the calculus is necessarily rough, this is the rationale of determining standards to reduce pollution of all kinds.

Just as governmental intervention is needed to bring about the reallocations of resources needed for environmental improvement, so it is also required to levy the costs of such improvement equitably among individuals and groups in society so as to improve—or at least prevent a worsening of—the distribution of income.

There are opposite approaches to the problem of cost allocation. By one principle, polluters should pay the costs of suppressing their pollution; by another, the public should pay polluters to stop polluting. The second principle is defended on the ground that the public benefits from the reduction of pollution and should pay the costs of this benefit. Those who espouse this view hold that tax credits and public subsidies are the proper instruments of a policy for environmental betterment.

Libertarians usually favor this approach because of their preference for the "carrot" versus the "stick," and their belief that public boards often come under the domination of those they are supposed to regulate.

Advocates of the first principle argue, to the contrary, that society initiates an anti-pollution policy from a current status of inequity. The problem is to restore equity as between polluters and those damaged by pollution, not to compensate polluters for a loss of equitable rights. They also observe that persons with large incomes generally generate disproportionately more pollution than those with low incomes, so that a policy of internalizing costs in the polluter will tend to shift income from richer to poorer people, with resulting gains in social well-being. The appropriate instruments for dealing with pollution are, in their view, public regulations to reduce harmful activities, or taxes and fines on polluters.

Equity requires that the costs of suppressing environmental damage be borne by those responsible for it. Public restraint of private actions harmful to the environment thus should be the dominant instrument of environmental policy. Assertion of this principle does not, however, preclude the use of taxes, fines, or lawsuits, nor does it rule out the use of public subsidies to enterprises which, through long-continued tolerance of harmful activities vital to their survival, have acquired a certain equity in them. For example, a city council might prohibit billboard advertising of off-premise goods or services, on the ground that the visual pollution costs borne by the public exceed the benefits. To enable outdoor advertising companies to finance an adjustment into other activities, a city might reasonably offer to pay them subsidies over a period of years on a descending scale.

Since the quality of the urban environment is a function of many variables, public policies to enhance the environment must utilize many instruments.

Direct governmental control of emissions of pollutants—audial, atmospheric, olfactory, visual, or health-affecting—is now exemplified in federal and state laws governing air and water pollution, and in federal standards of noise emissions from aircraft engines. Assuming that reduction of emissions is the least costly solution, the main problems are to determine appropriate standards and enforce them. In fixing standards, the state of pollution-control technology is an important consideration. Where such technology exists and can be applied at reasonable cost, the law should simply ban emissions and enforce compliance. This appears to be true of much air and water pollution from fixed sources, such as the chimneys of manufacturing and power-generating plants. Where pollution technology is in process of development, as in the case of automobile emissions, government should fix standards that are progressively raised as time goes on.

Another way to internalize external costs is to guarantee each property owner legal rights to the amenities pertaining to his property. A California court recently awarded substantial damages to home owners near the Los Angeles International Airport to compensate them for demonstrated loss of property values because of excessive noise from airplanes. A constitutional amendment should be enacted guaranteeing every property owner a right to environmental amenities, because this would induce business enterprises to reduce or eliminate pollution in order to escape legal liabilities. However, judicial processes are so costly, time-consuming, and uneven in their results as to make other solutions to environmental problems preferable.

Governments—federal, state, and local—themselves contribute to air and water pollution, especially by discharging untreated sewage into rivers and lakes. They should internalize these costs by massive public expenditures on sewage-treatment and water-purification plants. Such outlays will, of course, ultimately be paid for by a public that presumably values a clean environment more highly than the money paid in taxes to finance such facilities.

Urban planning, zoning, and building regulations are powerful instruments for enhancing the amenities of space, privacy, recreation, housing, transportation, and beauty in our cities. If American cities are to offer ample amenities for living, much stronger governmental controls of the design, quality, height, and density of buildings, and of the layout of transportation, recreation, and cultural facilities will be necessary. Americans will have to put a much higher priority on urban amenities, if strong enough instruments of social control over property usage are to be forged. Such controls will be opposed by builders, accustomed as they are to permissive public regulation that can be bent to their purposes. Yet firm public control of land usage under a long-range metropolitan plan is one reason why such cities as London hold a strong attraction for their residents as well as for millions of foreign visitors.

Enlargement of the supply of urban amenities also calls for immense public and private expenditures on recreational and cultural facilities, housing, and public transportation systems. The many programs coming under the auspices of the federal Departments of Transportation and of Housing and Urban Development serve this end. A whole battery of incentives for the participation of private enterprise in the gargantuan tasks will need to be fabricated, including tax credits, accelerated depreciation, credit guaranties, cost-plus contracts, and direct governmental subsidies. The naive idea that private corporations can or will undertake urban rehabilitation out of a sense of "social responsibility" denies the ineluctable fact that in a competitive market economy the firm cannot devote a material part of its resources to unprofitable activities and survive. Just as government must first create a market for pollution-reducing devices before the enterprise system will produce them, so it must first create adequate incentives to induce enterprises to produce urban housing and transit systems. That the responses are swift when the incentives are strong is shown by the great strength of the housing boom after World War II, triggered by liberal F.H.A. mortgage insurance and Veterans Administration home-loan guaranties.

Above all, a high-quality urban environment requires the public to assign high values to urban amenities—to appreciate them greatly and to work hard and pay for them. So far, too few American urbanites have held such values with sufficient intensity to bring about the necessary political action. Whether recent public outcries for a better environment will be sufficiently strong, sustained, and widespread to change the historical American posture of indifference remains to be seen.

The sudden emergence of the environmental problem raises profound issues about the functioning of our social institutions. Does it betoken an institutional breakdown—a failure to respond to new demands of the public? Has the social system responded, but been seriously laggard in its responses? Does the fault lie mainly in the political or in the market subsystem of our society, wherein there are two methods by which social choices of the uses of resources are made—voting in elections and buying in markets?

Although these questions cannot be answered finally, the most defensible positions appear to be the following. First, the social system has been sluggish in responding to the higher values placed by the public on environmental amenities, but it has not broken down and the processes of resource reallocation have begun. Second, the environmental crisis was generated primarily by tardy responses of the political system, and only secondarily by faults in the market system.

If American society is to attain optimal well-being, its dual set of political and market controls must operate promptly and in the proper sequence in response to changes in social values. Political action is first needed to create a demand for environment-improving products; market competition can then assure that this demand is satisfied economically. Measures are needed to improve both political and market processes.

Our model of the dynamic relationships between changes in social values, government actions, and corporate behavior is shown in the chart below. The primary sequential flow of influence runs from changes in social values, via the political process, to changes in governmental regulation of the private sector and reallocation of public resources;

Dynamic Relationships Between Public Values, Governmental Regulations, and Corporate Resource Allocations

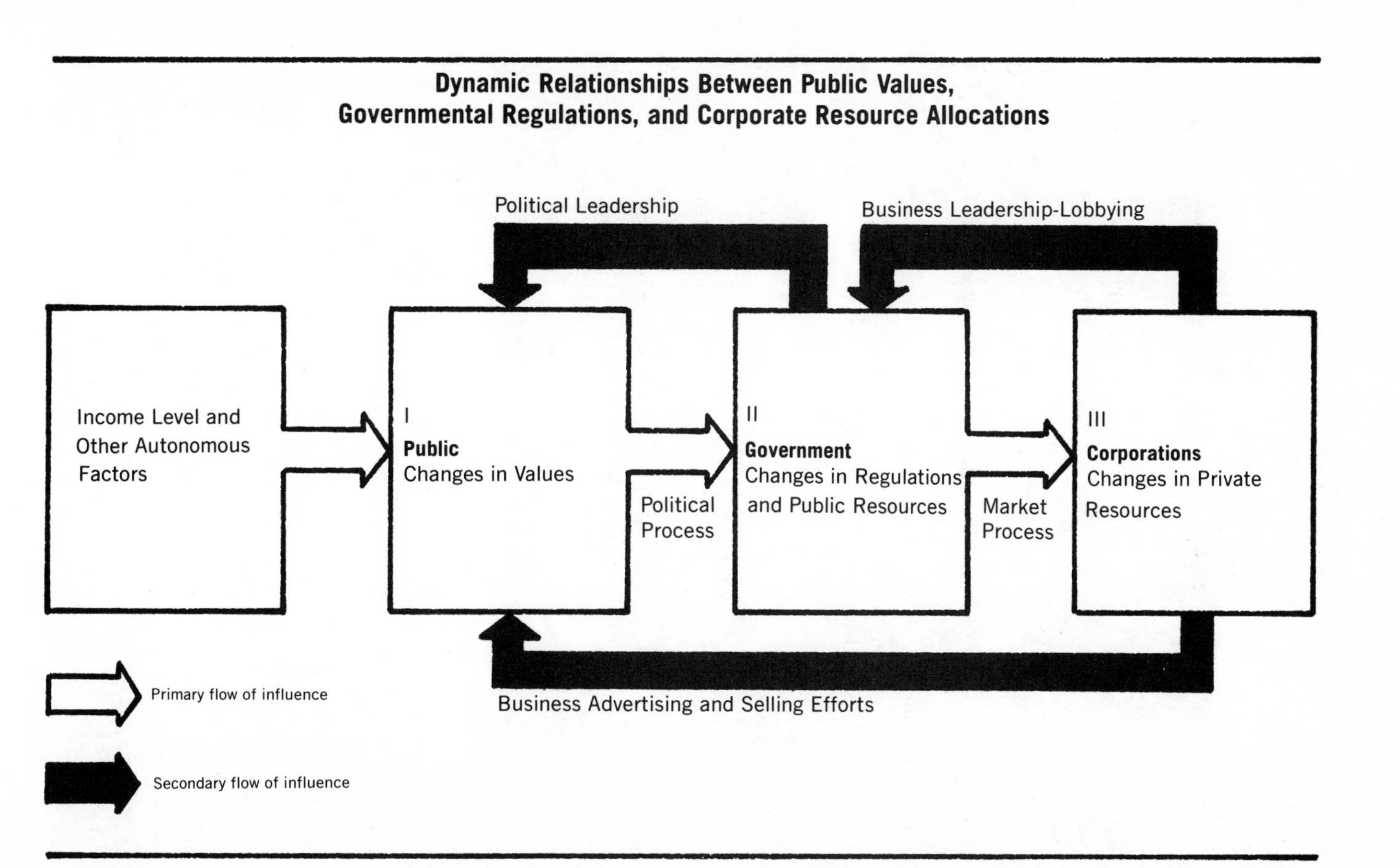

thence, via the market process, to corporate reallocation of private resources. However, changes in social values are not wholly determined by shifts in levels of income and other autonomous factors. They also respond to political leadership in the legislative and executive branches of government and to the public advertising and selling efforts of corporations. Similarly, governmental actions are not responsible exclusively to shifts in the values of the public. They are also influenced in some degree by the political activities of businessmen and by corporate lobbying. These secondary flows of influence also help to determine the performance of the social system.

The model enables us to identify salient points of improvement in the system. They are to reform the political process so that government actions will more rapidly and accurately reflect significant shifts in social values, to reform the market process so that corporate behavior will more rapidly and accurately reflect changes in governmental regulation, and to reform political and business behavior so that their secondary influences will help rather than hinder. Specifically, what changes are needed in each of these three areas?

The environmental problem emphasizes once again the need for a political system capable of translating changes in social values rapidly and accurately into governmental actions. The political apparatus for sensing, recording, mobilizing, transmitting, and acting upon millions of changes in individual preferences must be improved. Our representative system of government must be made more representative. This raises anew the old dilemmas of participative democracy, the weaknesses of political parties, the unrepresentativeness of legislatures, and the inordinate influence of pressure groups in an age of accelerating technological, demographic, and economic change. The basic requirements for greater efficiency of the political system are better education and sustained participation in political affairs by citizens. While one may easily be pessimistic in the light of the past, there is ground for hope of improvement. Americans generally spend only a small fraction of the time and effort they devote to private goods in making choices of the public goods they purchase with their taxes. Yet purchases of public goods and services are now nearly one-third as large as purchases of private goods. During 1969 government purchases amounted to $225 billions or twenty-three per cent of a total G.N.P. of $923 billions, whereas private purchases were $698 billions or seventy-seven per cent. Rational behavior in resource allocation requires a massive increase in the time and effort devoted to public decisions. Hopefully, the present egregious under-allocation of time represents a cultural lag which will be rectified in due course.

Changes should be made in the voting process to make it function more like a market. Just as consumers record the relative intensities of their demands for different private goods by the amounts of their expenditures in markets, so voters could be enabled to record the relative intensities of their demands for public goods. Each voter could be given, say, one thousand votes, which he could cast in whatever numbers he chose for alternative aggregate levels of public expenditures and alternative patterns of allocation of each level among different objects. Finally, a maximum usage of direct links between public expenditures and the taxes levied to finance them, could help to make the political system more responsive.

The competitive market system must also be made more responsive to shifts in public values and governmental regulations. Despite its undoubted superiority as a device for gauging consumers' wants, the recent rise of the "consumerism" movement reflects, among other things, a disturbing insensitivity of the business corporations to changing public demands and expectations. The foot-dragging behavior of the auto makers in regard to safety and air pollution and of the oil companies in regard to air and ocean-water pollution are symptomatic. Business corporations generally have been reluctant, if not obstructive, reactors to new social values instead of innovative leaders in satisfying them. Either their market researchers have been unable to detect them, or else correct market intelligence has not been utilized by their engineering, manufacturing, and marketing executives.

A reorientation of corporate organization is needed, from the board of directors down through corporate and divisional managers to individual plant and store executives. The board should include one or more "outside" directors chosen especially for their knowledge of corporate relationships to society, including the environment. This need should be met by the normal process of including such nominees in the slate of directors presented by management for stockholder vote, rather than by augmenting the board by special stockholder nominees, as Ralph Nader proposed to General Motors Corporation. The normal procedure is more likely to result in effective board action to improve the environment, because it avoids "bloc" politics within the board. Every single policy and action of the firm should be reviewed for its effect upon the environment. An environmental analyst, assigned to this task as a staff adviser to the chief executive, would help to assure good corporate behavior. Standard corporate policy should require all managers to include in their proposals for new operations of facilities measures for preventing adverse environmental effects. Corporations should also make more penetrating use of consumer surveys and public-opinion polls in order to keep informed of shifts in public tastes and priorities.

Reforms are also needed to insure that the secondary influences upon social values exercised by political and business leaders are facilitative rather than obstructive. These influences are significant. For example, President Eisenhower's sponsorship of the Interstate Highway Act in 1956 and President Kennedy's proposal of a manned round trip to the moon in 1961 mobilized and activated changes in the values of the American people which led to highway and space programs each of the order of five billion dollars a year. President Nixon's leadership in 1970 in a national effort to improve the environment will probably produce even larger reallocations of resources. All three Presidents discerned deep changes in public priorities to which they gave form and implementation. Without such political leadership, readjustments would have been delayed amid mounting public tension and frustration.

American corporate leadership generally has not played a helpful role in implementing changes in social values. Whereas business lobbyists should be informing legislators of new environmental regulations desired by the public, they usually oppose such changes. Most corporate advertising is narrowly focused upon expanding public demand for existing products rather than for new products with superior environmental effects. As Henry Ford recently advised, corporate managers should "stop thinking about changing public expectations as new costs which may have to be accepted but certainly have to be minimized. Instead, we should start thinking about changes in public values as opportunities to profit by serving new needs."

This analysis of deterioration in the urban environment and of means to restore it has unveiled neither a master culprit nor a panacea. It has delineated a complex public problem requiring many instruments of policy for its solution. It has shown that the basic requirement is a citizenry that assigns higher values to urban amenities than it has in the past, and will work harder and pay more to get them. Given new social preferences, new regulations will be imposed and those long-neglected regulations on the statute books will finally be enforced. It is disturbing to reflect that a lawsuit brought by the Attorney General of the United States early in 1970 against several large corporations for polluting the southern end of Lake Michigan was to enforce a federal statute enacted in 1899. Here—as in the administration of urban zoning codes—Americans have not put high enough values upon environmental amenities to insist that private actions conform to existing public laws.

Environmental improvement will call for annual public and private expenditures of tens of billions of dollars indefinitely into the future. Profound changes will be necessary in the structure of relative costs and prices of goods, and in patterns of production and consumption. These readjustments will cause difficulties for individual companies operating on the margin of profitability and unable to pay the full costs of their products. Yet the ability of our profit-oriented enterprise system to adapt to a massive internalization of costs cannot be doubted, when one recalls its successful assimilation of the technological revolution since World War II. Over a period of time the costs and prices of products with large external costs (e.g., automobiles) would rise relatively, while those with large external benefits (e.g., homes) would decline relatively. While consumers would spend relatively less on autos and relatively more on housing, in a growing economy this would mean changes in the growth rates of different industries rather than an absolute decline in the output of any one. Also, new industries would emerge to supply the growing demand for pollution-controlling equipment and services. Profit rates and market signals would continue to guide resources in the directions desired by consumers.

The effects of environmental improvement upon the overall growth of the U.S. economy depend mainly upon how "economic growth" is defined and measured. There is a growing recognition that the true end of public policy is a steady expansion of social well-being, and that a rising G.N.P. is only a means to this end. G.N.P. is simply a measure of the aggregate output of the economy, whereas social well-being is also directly related to the composition of output, its full costs, and the uses to which it is put. If, as has been true during the past twenty years, much production included in the G.N.P. has been associated with national defense and environmental degradation, growth of the G.N.P. can be a highly misleading index of gains in social well-being. Indices of well-being should be developed to help guide long-term public policy, and G.N.P. also should be recast to provide a more meaningful measure of total output.

Assuming the existence of a strong effective demand by the public for a better urban environment, it cannot be doubted that a redirection of production to supply that demand will expand the well-being of American society. A better environment would enable people to reduce many other costs they now incur for health, property maintenance, recreation, and travel to leave uncongenial surroundings. Rising social well-being is not in conflict with an expanding G.N.P., provided that the increments of production improve the quality of life. On the contrary, growth of production is needed for that purpose. As President Nixon said in his 1970 State of the Union Message: "The answer is not to abandon growth, but to redirect it."

NOTES ON THE AUTHORS

Peter Andrews is an affiliated student in archeology and anthropology, St. John's College, Cambridge, England.

Aristotle (383-322 B.C.) was a Greek philosopher whose *Politics* was and is still one of the most widely read books on government.

Joseph Connell is a member of the biology department at the University of California at Santa Barbara.

Charles Dickens (1812-1870), English novelist, often described the horrifying conditions in English industrial areas in the nineteenth century.

René Dubos, professor at Rockefeller University, is a microbiologist and pathologist who has won many awards for his scientific contributions and has written several books on science and health.

Paul R. Ehrlich is a biologist on the faculty at Stanford University and an Associate of the Center for the Study of Democratic Institutions.

John Fischer, a contributing editor of *Harper's Magazine,* writes a monthly column on current affairs called "The Easy Chair."

Anatole France (1844-1924) was a French novelist and critic of modern society. His *Penguin Island* is a satire on French politics, religion, and art.

Louis J. Fuller, Air Pollution Control Officer for Los Angeles County, has won worldwide attention for his efforts to overcome pollution from automobiles.

A. J. Haagen-Smit, a scientist at the California Institute of Technology, has become well-known for his research into the causes and solution of air pollution.

Sir Julian Huxley, zoology professor, philosopher, and writer, has been Director General of UNESCO and has won awards for his scientific writings.

Hugh Iltis is a professor of botany at the University of Wisconsin, Madison.

Neil Jacoby, an economist at the University of California at Los Angeles and an Associate of the Center for the Study of Democratic Institutions, recently headed a presidential task force.

Joseph Wood Krutch was drama critic for *The Nation* from 1924 to 1950. He is the author of many books on the theater, eighteenth-century literature, and nature. Among the last are *The Desert Year, The Measure of Man, Grand Canyon,* and *The Forgotten Peninsula.*

Daniel Lang is a journalist with a special interest in problems of nuclear science. His books include *Early Tales of the Atomic Age, The Man in the Thick Lead Suit,* and *From Hiroshima to the Moon.*

John Lear is Science Editor of *The Saturday Review.*

Orie Loucks is a professor of botany at the University of Wisconsin, Madison.

Thomas Robert Malthus (1766-1834), English economist, is best known for his *Essay on the Principle of Population as it affects the Future Improvement of Society, with Remarks on the Speculations of Mr. Godwin, M. Condorcet, and other Writers,* the enlarged and amended version of which appears in this book.

Gene Marine has worked in magazine journalism and broadcasting and was a writer for *The Nation* and *Ramparts.*

André Maurois, the noted French author and humanist, is the author of many novels and biographies. His works include *The Silences of Colonel Bramble; Proust, Portrait of a Genius; Lelia, or the Life of George Sand; September Roses; A History of France;* and *The Titans.*

Jean Mayer is professor of nutrition, lecturer on the history of public health, and a member of the Center for Population Studies at Harvard University. He is an authority on nutritional problems in the United States and underdeveloped countries.

Margaret Mead, anthropologist and curator of ethnology at the American Museum of Natural History, is the author of *Coming of Age in Samoa* and *Male and Female.*

John Stuart Mill (1806-1873) was a political philosopher whose books *On Liberty* and *Principles of Political Economy* predicted many of the problems that have arisen in the modern world.

Peter Millones is a consumer reporter for *The New York Times.*

Lewis Mumford, considered an outstanding authority on architecture and urban affairs, is the author of many books including *The Urban Prospect, The Cult of Cities,* and *The Conduct of Life.*

William Murdoch is a member of the biology department at the University of California at Santa Barbara.

Neil Postman is a professor of education at New York University. He is co-author of *Education as a Subversive Activity.*

Frank M. Potter, Jr., is executive director of Environmental Clearinghouse, Inc.

Peter Schrag is Editor-at-Large for *The Saturday Review.*

Jonathan Swift (1667-1745), an Irish clergyman and satirist, wrote *Gulliver's Travels, The Battle of the Books,* and many political essays.

Allan Temko has written on architecture and urban problems for the *San Francisco Chronicle,* the *Washington Post, Harper's,* and other periodicals.

Henry David Thoreau (1817-1862) was a philosopher and nature lover whose books include *Walden, or Life in the Woods* and *A Week on the Concord and Merrimack Rivers.*

Arnold J. Toynbee is an eminent British historian whose major work is *A Study of History.*

Stewart L. Udall is a former Secretary of the Interior and a dedicated conservationist.

Edward Weeks writes a regular column, "The Peripatetic Reviewer," for *The Atlantic.*

George Whicher (1889-1954), professor and critic, was the author of several books on American literature.

Leonard Wolf is a writer professionally engaged in studies of environmental health.